Introduction
to Relativity

Complementary Science Series

Earth Magnetism

Wallace Hall Campbell

Physics in Biology and Medicine, 2nd Edition

Paul Davidovits

Fundamentals of Quantum Mechanics

J.E. House

Chemistry Connections

Kerry Karukstis ▶ *Gerald Van Hecke*

Mathematics for Physical Chemistry, 2nd Edition

Robert Mortimer

The Physical Basis of Chemistry, 2nd Edition

Warren S. Warren

www.harcourt-ap.com

Introduction to Relativity

John B. Kogut

Department of Physics
University of Illinois at Urbana-Champaign
Urbana, Illinois

A Harcourt Science and Technology Company

San Diego   San Francisco   New York   Boston   London   Sydney   Tokyo

Sponsoring Editor	Jeremy Hayhurst
Production Managers	Joanna Dinsmore and Andre Cuello
Editorial Coordinator	Nora Donaghy
Marketing Specialist	Stephanie Stevens
Cover Design	G.B.D. Smith
Copyeditor	Julie Nemer
Composition	Technical Typesetting Incorporated
Printer	Sheridan Books

Series Logo: Images © 2000 PhotoDisc, Inc.

This book is printed on acid-free paper. ∞

Academic Press
A Harcourt Science and Technology Company
525 B Street, Suite 1900, San Diego, California 92101-4495, U.S.A.
http://www.academicpress.com

Academic Press
Harcourt Place, 32 Jamestown Road, London NW1 7BY, UK
http://www.academicpress.com

Harcourt/Academic Press
A Harcourt Science and Technology Company
200 Wheeler Road, Burlington, Massachusetts 01803
http://www.harcourt-ap.com

Library of Congress Catalog Card Number: 00-111388

International Standard Book Number: 0-12-417561-9

PRINTED IN THE UNITED STATES OF AMERICA
01   02   03   04   05   06   SB   9   8   7   6   5   4   3   2   1

▶

Contents

▶

Preface

Overview—What Can You Expect from This Book?

Before we get to the real beginning of this book, let's look at its aims and objectives. Its intention is to teach you how to think about relativity and understand it in the simplest possible way. There are not many formulas, algebra is kept to a minimum, arguments are short and to the point, and calculus is *not* allowed until the last chapter. Vague, pseudophilosophical terms are outlawed. The inspiration for this approach is, naturally, Einstein himself. Physicists love to read Einstein's original papers because they typically included "thought experiments" that illustrated the concepts under investigation in a fashion even a child (a very smart one, anyway) could understand. For example, in this book we construct a simple clock out of mirrors and a light beam, following one of Einstein's most famous arguments, and we see how it works in a frame where it is at rest and in a frame in which it moves. In this way, we derive both time dilation (moving clocks run slower than clocks at rest) and length contraction (moving rods are shorter than rods at rest) in a transparent way.

In the course of these arguments, it will become apparent that at the heart of these unfamiliar phenomena is the fact that clocks that are separated in space and synchronized in one frame are *not* synchronized in a frame in relative motion. This fact, often called the relativity of simultaneity, is the essence of the subject. It follows simply from the basic experimental fact that there is a speed limit in nature that must be common to all observers whether they are in relative motion or not. A physical realization of the speed limit is afforded by light. We can synchronize spatially separated clocks by sending light rays between them. In this way, one of the basic postulates of relativity, that there exists a common speed limit in all inertial frames, dictates the operation and synchronization of all our clocks.

The content of relativity lies in its two basic postulates:

1. The laws of physics are the same in all inertial frames of reference.

2. There is a common finite speed limit in all inertial frames.

I hope that the novice reader is intrigued by all this. I also hope that the more seasoned physicist/teacher appreciates that we are going to build the subject up from the ground level. We are *not* going to follow a historic approach where electrodynamics played a central role in the development of the subject. That approach fails the novice. Instead, just from the idea that any inertial frame is as good as any other and the fact that there is a speed limit common to all frames, we can derive all the classic results of relativity without any specialized mathematics. The ideas of invariant intervals, even the Lorentz transformations, which are typically so prominent in relativity textbooks, are relegated to a later chapter where they are introduced as conveniences that summarize what we have already figured out through simple arguments. Visualization plays a central role in these developments. We introduce Minkowski space–time diagrams, which show the spatial and temporal coordinates of two frames in relative motion **v**. Time dilation, Lorenz contraction, and, most important, the relativity of simultaneity (spatially separated clocks that are at rest and synchronized in one frame are not synchronized in a frame in relative motion) can be understood through simple pictures (cartoons, really), both qualitatively and quantitatively. From this perspective, we study the Twin Paradox and see that there is nothing really paradoxical about it—when one twin leaves the other and takes a round trip, she returns younger than her sibling. Of course, it is one thing to understand a physical phenomenon and another thing to feel comfortable with it. I know of no one who feels comfortable with the Twin Paradox, although high-energy experimentalists observe the effect daily in their experiments with high-energy elementary particles. Anyway, I hope that the reader can understand it and see that there is no contradiction involved in it.

Once we have mastered kinematics—the measurement of space and time intervals—we pass on to introductory dynamics, the study of energy, momentum, and equations of motion. In Newtonian mechanics we recall that there is momentum conservation, on the one hand, and mass conservation, on the other. In relativity, however, once we have figured out how spatial and temporal measurements are related in inertial frames, the definition and properties of momentum, energy, and mass are determined. We present another of Einstein's famous thought experiments that shows that energy and inertia (mass) are two aspects of one concept, relativistic energy, and find $E = mc^2$. The properties of relativistic momentum then follow from the properties of spatial and temporal measurements.

Although it is conventional to understand Postulate 2 (kinematics) thoroughly before delving into Postulate 1 (dynamics), the consistency of the two is central. In particular, Einstein asserted Postulate 1: If we have an inertial frame of reference in which isolated particles move at constant velocities in straight lines, then any frame moving with respect to us at velocity **v** has the same properties and there is no physical distinction between the two frames (i.e., the laws of physics are the same in both frames). This postulate constrains how objects interact. The dynamical laws of physics must then be consistent with Postulate 2, which states that there is a universal speed limit.

Newton's laws of dynamics are not consistent with the existence of a speed limit. We have to modify the notions of momentum, energy, and mass and the law of acceleration in the presence of a force to invent dynamical rules that are relativistic (i.e., that are consistent and lead to Postulate 2, the existence of the speed limit). We do this aided, again, by the original thought experiments of the masters, Einstein and Max Born. In this case, a close look at an inelastic collision in several frames of reference leads us to relativistic momenta and energy. From there we obtain the relativistic version of Newton's Second Law, that mass times acceleration equals force. We then consider high-energy collisions between particles and illustrate how energy can be converted into mass and how mass can be converted into energy in our relativistic world.

Our last topic consists of an introduction to the General Theory of Relativity, where we finally consider accelerated reference frames in Einstein's world. The key insight here is Einstein's version of the Equivalence Principle: There is no physical means to distinguish a uniform gravitational field from an accelerated reference frame.

This principle has many interesting forms and applications. Suppose you are on the surface of Earth and want to understand the influence of the gravitational field on your measuring sticks and clocks compared to those of your assistant who is at a greater height in the Empire State Building. Einstein suggests that the assistant jump out the window, because in a freely falling frame all effects of gravity are eliminated and we have a perfectly inertial environment where special relativity holds to arbitrary precision! During his descent, your assistant can make measurements of clocks and meter sticks fixed at various heights along the building and measure how their operation depends on their gravitational potential. We pursue ideas like this one in the book to derive the gravitational red-shift, the fact that clocks close to stars run more slowly than those far away from stars; the resolution of the Twin Paradox as a problem in accelerating reference frames; and the bending of light by gravitational fields.

A fascinating aspect of the Equivalence Principle is its universality, which becomes particularly clear when we calculate the bending of light as a ray glances by the Sun. Why does the light ray feel the presence of the mass of the Sun? The Equivalence Principle states that an environment with a gravitational field is equivalent to an environment in an accelerating reference frame—explicit acceleration clearly affects the trajectory of light and any other physical phenomena. So gravity becomes a problem in accelerated reference frames, which is just a problem in coordinate transformations, which is an aspect of geometry! We show that this problem in geometry must be done in the context of four-dimensional space–time, our world of Minkowski diagrams. Einstein's theory of gravity brought modern geometry, the study of curved spaces, into physics forever. We see hints of this in our work here, but our discussions do not use any mathematics beyond algebra and an elementary integral or two.

As long as we concentrate on gravitational fields of ordinary strength, we are able to use the Equivalence Principle to make reliable, accurate predictions. The Equivalence Principle reduces gravity to an apparent force, much like the centrifugal or Coriolis forces that we feel when riding on a carousel. In fact, we use relativistic turntables and rotating reference frames as an aide to studying and deriving relativistic gravitational effects. A short excursion into the world of curved surfaces helps us appreciate general relativity in applications where the gravitational field varies from point to point. We end our discussion with a look at current puzzles and unsolved problems.

Although the perspective of this book is its own, it owes much to other presentations. The influence of C. P. French's 1968 book [1] *Special Relativity* is considerable, and references throughout the text indicate where my discussions follow his. To my knowledge, this is the finest textbook written on the subject because it balances theory and experiment perfectly. The reader will find discussions of the Michelson–Morley experiment and early tests of relativity there. I do not cover those topics here because this book is aimed at students who may know little electricity and magnetism at this stage of their education. French's discussions of energy and momentum are reflected in my later chapters, and his problem sets are a significant influence on those included here. The exposition by N. D. Mermin [2] influenced several discussions of the paradoxes of relativity. This book is also recommended to the student because it shows a condensed-matter physicist learning the subject and finding a comfort level in it through thought-provoking analyses that avoid lengthy algebraic developments. The huge book by J. A. Wheeler and E. F. Taylor [3] titled *Spacetime Physics* inspired several of our discussions and problem sets. This book, a work that only the unique, creative soul of John Archibald Wheeler could produce, is recommended for its leisurely, interactive, thought-provoking character. Finally, the books by W. Rindler [4, 5], a pioneer in modern general relativity, are most highly recommended. His book *Essential Relativity* [5] is my favorite introduction to general relativity. After the student has mastered electricity and magnetism and Lagrangian mechanics, he or she could read *Essential Relativity* and see how the introductory remarks on general relativity in the second half of this book can be made quantitative.

My thanks to Mirjam Cvetic of the University of Pennsylvania, Charles Dyer of the University of Toronto, David Hogg of Princeton University, and Richard Matzner of the University of Texas–Austin for their helpful comments on the manuscript.

John B. Kogut

Chapter 1

Physics According to Newton—A World with No Speed Limit

When you set up a problem in Newtonian mechanics, you choose a reference frame. This means that you set up a three-dimensional coordinate system, for example, so any point $\mathbf{r}$ can be labelled with an x measurement of length, a y measurement, and a z measurement, $\mathbf{r} = (x, y, z)$. Cartesian coordinates might not be the most convenient in a particular case, so you might use spherical coordinates or whatever suits the problem best.

Newton imagined carrying out experiments on a mass point m in this coordinate system. He imagined that the coordinate system was far from any external influences and under those conditions he claimed, on the basis of the experiments of Galileo and others, that the mass point could only move in a straight line at a constant velocity. Newton labeled such a frame of reference "inertial." Next, if the mass point were subject to a force, Newton claimed that its velocity would change according to his Second Law, force equals mass times acceleration,

$$\mathbf{f} = m\mathbf{a}. \tag{1.1}$$

The mass m in Eq. (1.1) is clarified in part by the Third Law, which states that if a body exerts a force $\mathbf{f}_1$ on another body, then the second body exerts a force $\mathbf{f}_2 = -\mathbf{f}_1$ on the first. This postulate is called the Law of Action–Reaction and it implies in this case that

$$m_1\mathbf{a}_1 = -m_2\mathbf{a}_2. \tag{1.2}$$

So, if the first body sets the scale for inertia, in other words, we define $m_1 \equiv 1$, then m_2 follows from Eq. (1.2).

Newton and others realized that there must be a wide class of inertial frames. If we discovered one frame that was inertial, then Newton argued that other inertial reference frames could be generated by

1. **Translation**—move the coordinate system to a new origin and use that system.

2. **Rotation**—rotate the coordinate system about some axis to a fixed, new orientation.

3. **Boosts**—consider a frame moving at velocity **v** with respect to the first.

Properties 1 and 2 are referred to as the uniformity and isotropy of space. Property 3, referred to as Galilean invariance, is plausible because if we have two frames in relative motion at a constant linear velocity **v**, then acceleration measurements are the same in both frames. So, Eqs. (1.1) and (1.2) are unaffected—the Galilean boost has no physical impact on the dynamics and so should be a symmetry of the theory.

The reader should be aware that Newton and his colleagues argued constantly about these points. What is the origin of inertia? Why are inertial frames so special? Can't we generalize the Properties 1–3 to a wider class of reference frames? We will not discuss these issues here, but the reader might want to pursue them for a greater historical perspective. Our approach is complementary to the traditional one and carries less intellectual baggage.

Underlying Newtonian mechanics are the concepts of space and time. To measure the distance between points we imagine a measuring rod. Starting from an origin $(0, 0, 0)$, we lay down markers in the x and y and z directions so we can measure a particle's position $\mathbf{r} = (x, y, z)$ in this inertial frame. Next we need a clock. A simple device like a simple harmonic oscillator will do. Take a mass point m on the end of a spring and let it execute periodic motion back and forth. Make a convention that one unit of time passes when the mass point goes through one cycle of motion. In this way we construct a clock and measure speeds of other masses by noting how far they move in several units of time—in other words we compare the motion of our standard mass point in our simple harmonic oscillator with that of the other mass. Note that this is just what we mean by time in day-to-day situations. To make an accurate clock we need one with a sufficiently short unit of time, or period. Clocks based on the inner workings of the atom can be used in demanding, modern circumstances.

Just as it was convenient to place distance markers along the three spatial axes $\mathbf{r} = (x, y, z)$, it is convenient to place clocks on the spatial grid work. This will make it easy to measure velocities of moving particles—we just record the positions and times of the moving particle on our grid of measuring rods and clocks. When the moving particle is at the position **r** and the clock there reads a time t, we record the event. Measuring two such events allows us to calculate the particle's velocity in this frame if the particle has a uniform velocity. In setting up the grid of clocks, we must synchronize them so we can obtain meaningful time differences. This is easy to do. Place a clock at the origin and one at $\mathbf{r} = (1, 0, 0)$. Then, at the halfway point between them, place a beacon that sends out a signal in all directions. Because space is homogeneous and isotropic, the signal travels at the same speed toward both clocks. Set the clocks to zero, say, when they both receive the signal. The two clocks are synchronized and we didn't even need to know the speed of the signal emitted from the beacon. Clearly we could use this method to synchronize all the clocks on the grid.

Now we are ready to do experiments involving space-time measurements in this frame of reference. Call this frame S. It consists of the gridwork of measuring rods and clocks, all at rest, with respect to each other. Now suppose we want to compare our experimental results with those obtained by a friend of ours at rest in another fame that moves at velocity $\mathbf{v} = (v_x, 0, 0)$ with respect to us. According to our postulates, his or her measurements are as good as ours and all our physical laws can be written in his or her frame of reference without any change (Figure 1.1).

The measuring rods in S′ are identical to those in S; the clocks in S′ are also identical to those in S. Suppose that the origin of S and S′ coincide when the clock at the origin in S reads time $t = 0$ and the clock at the origin in S′ reads $t' = 0$. Now for the crucial question: Do the other grid markings and clocks at those markings also agree in the two frames S and S′? There are two distinct physics issues to consider here. The first is the operation of the clocks and rods in each frame. Following Postulate 1, all the clocks and rods in S work exactly the same as those at rest in S′. We need only

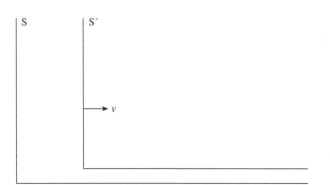

FIGURE 1.1 ▶

know that both frames are inertial—the relative velocity between the frames is physically irrelevant to the dynamics within each. The second issue is the physical mechanism by which we can transmit the information in one grid of rods and clocks to the other at relative velocity **v**. It follows from Newton's Second Law that information can be transmitted instantaneously from one frame to another. In Newton's world, objects can be accelerated to unbounded relative velocities. Accordingly, signals and information can be transmitted at unbounded velocities, essentially instantaneously. Therefore, the spatial gridwork and the times on each clock in S′ can be instantaneously broadcast to the corresponding rods and clocks in the frame S. Therefore, the gridwork of coordinates and times in both frames must be identical, even though they are in relative motion. So, the lengths of measuring rods and the rates of clocks are independent of their relative velocities. We therefore need only one measuring rod and one clock at one point in one inertial frame, and we know the positions and times of all measuring rods and clocks in any other inertial frame. For the purposes of this book, this serves as the meaning of "absolute space" and "absolute time" in Newtonian mechanics. We need not hypothesize about these notions from some philosophical basis, as was done historically, but can deduce them from the fact that in Newtonian mechanics there is no speed limit—information can be transmitted instantaneously.

We have dealt with these issues very explicitly because they will help us appreciate special relativity, where there is a speed limit, the speed of light. In all the rules of Newtonian mechanics, the rules of how things work, the Second and Third Laws, do not distinguish between inertial frames. The dynamics do satisfy a principle of relativity. The difference between the two theories comes from the fact that one has a speed limit and one does not. This affects how information is shared between frames. It also affects the dynamics within each frame—Einstein's form of Newton's force equals mass times acceleration is different because it must not permit velocities greater than the speed limit. Each theory is consistent within its own rules. For example, the rule by which times and position measurements are compared between the two frames, S and S′, in relative motion in Newton's world reads

$$
\begin{aligned}
x &= x' + vt \\
y &= y' \\
z &= z' \\
t &= t',
\end{aligned}
\tag{1.3}
$$

where x, y, z, and t are measurements in S and x', y', z', and t' are the corresponding measurements in S′. For example, if we measure the position of a particle in S′ to be x', y', and z', at time t', then the coordinates of this event (measurement) in frame S are given by Eq. (1.3). These relations,

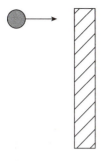

FIGURE 1.2 ▶

which are so familiar and obvious, are called Galilean transformations. Note that the first of them, $x = x' + vt$, states that the position of the particle at time t consists of two pieces: (1) the distance vt between the origins of the two frames at time t and (2) the distance x' from the origin to the particle in the frame S'. This rule contains the notion absolute space—it uses the fact that in a Newtonian world the distance x' in the frame S' is also measured as x' in the frame S. The last equation in Eq. (1.3) is the statement of absolute time. Note also that if the particle has a velocity v'_p with respect to the frame S' in the x' direction, then $x' = v'_p t' = v'_p t$, and its position in frame S is $x = (v'_p + v)t$, so its velocity relative to the origin of frame S is v_p,

$$v_p = v'_p + v, \tag{1.4}$$

which is the familiar rule called addition of velocities. It rests on Newton's ideas of absolute space and time. It will be interesting indeed to see how the results from Eqs. (1.3) and (1.4) are different in Einstein's world!

Perhaps it has not occurred to the reader before that information could be transmitted instantaneously in a Newtonian world. Actually, if the reader recalls some elementary problems in dynamics, the assumption was lurking just under the surface.

Suppose you consider the collision of a point particle and a rod, as shown in Figure 1.2.

When you solve this problem using Newtonian mechanics using conservation of linear and angular momentum, you implicitly assume that the impact at the upper edge of the rigid rod instantaneously accelerates the center of mass of the rod and begins rotating the bottom end of the rod to the left. If information was transmitted at a finite velocity, call it c, the lower edge of the rod could not know about the collision until a later time, $t = \ell/c$ where ℓ is the length of the rod. It is clear that the very notion of rigid bodies itself requires a theoretical basis where information is transmitted instantaneously. Rigid bodies cannot exist in Einstein's world. Of course, in practical problems where ℓ is a few meters, say, and c is enormous, $c \approx 3.0 \cdot 10^8$ m/s, the time delay encountered here is utterly negligible. (But it certainly would be significant if ℓ were the diameter of a star.) Nonetheless, the matter of principle is our major concern at the moment.

Another place where instantaneous transmission of information occurs in Newtonian mechanics problems is in the use of potentials. When we have two bodies interacting through a mutual force that is described through a potential that depends on the distance between the particles, we are assuming that the force on each particle is determined by the relative position of the other particle at that exact moment. No account is given of the transmission of that information between the separate particles. This Newtonian notion is called action at a distance and it underlies some of the greatest successes of nonrelativistic mechanics.

Chapter 2

Physics According to Einstein

2.1 A World with a Speed Limit

According to Newtonian mechanics you could place a charged particle (charge q, mass m) into a constant electric field E and accelerate it to an arbitrary velocity, $v(t) = (qE/m)t$. In addition, you could consider inertial frames in relative motion and the relative velocity could be arbitrarily large.

Unfortunately, nature does not allow such freedom. Modern accelerator experiments indicate that particles cannot be accelerated beyond a universal speed limit, which is the speed of light, $c = 2.9979\ldots \cdot 10^8$ m/s. Light is nature's fastest runner. Although historically the fact that the speed of light is the speed limit was extraordinarily important, it is not important for the logical development of the subject. Just the existence of a speed of light is all that really matters. We do not need to know anything about electromagnetism to derive time dilation and Lorentz contraction, and thereby overthrow Newton's vision of absolute space and time. But it is crucial to understand that the existence of a speed limit must be compatible with Postulate 1 of relativity, that the laws of physics are the same in all inertial frames. This means that all inertial frames must find the same universal value for the speed limit through their own experiments.

In particular, suppose that an experimenter at rest in frame S′ shown in Figure 2.1 turns on a flashlight and points it to the right. Those light rays can be measured as traveling at the speed limit c with respect to any observer at rest in frame S′. But an experimenter in frame S must also measure the speed of the light ray to be the speed limit c, independent of v! Just for fun, take an extreme example. Choose $v/c = 0.99999999$ and point the flashlight to the left, in the direction opposing the velocity v. An observer at rest in frame S′ measures the velocity of the light rays emanating from the flashlight and

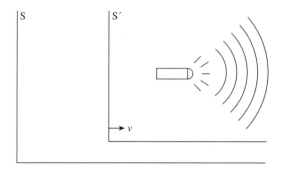

FIGURE 2.1 ▶

finds $c \approx 3 \cdot 10^8$ m/s to the left, as usual. But an observer at rest in frame S measures the velocity of light rays, too, and finds $c \approx 3 \cdot 10^8$ m/s to the left as well, instead of the value 3 m/s to the left that we would predict in Newton's world! The speed of the source of light, the flashlight, is utterly irrelevant! Both observers must measure the same speed limit because the laws of physics are identical in both of their frames. The Galilean result of Eq. (1.3), known as addition of velocities, which we have held as obvious, is just plain wrong. It will take us some time and effort to feel comfortable with the physics in Einstein's world!

In summary, the two postulates of Einstein's special relativity are

1. The laws of physics are the same in all inertial frames.

2. There is a speed limit.

The rest of this book will work out the implications of these two postulates. We begin by studying how measurements of rods and clocks are related between different inertial frames. Then we turn back to Postulate 1 and see how Newton's laws must be modified, how momentum and energy should be defined and related, in order to be compatible with the existence of a universal speed limit.

2.2   Making a Clock with Mirrors and Light

Just as in our discussion of Newtonian measurements of the positions and times of events, we set up a gridwork of coordinates and clocks in two frames, S and S′, which are both inertial and have a relative velocity v in the x direction, as depicted Fig 2.1. We synchronize clocks in frames S and S′ just as we described in the Newtonian world of Chapter 1. Our next task is to compare the positions and times on the clocks in S with those in S′.

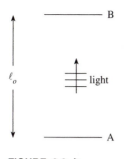

FIGURE 2.2 ▶

We need to construct a clock so that we can understand its operation using just Postulates 1 and 2. Einstein thought up a simple clock whose two ingredients, a basic length and a basic velocity, are easily understood in terms of Postulates 1 and 2. Take two mirrors at rest in S' and let them be separated by a distance ℓ_o in the y' direction. Both mirrors move relative to S in the x direction at velocity v and maintain their separation ℓ_o as measured in either frame. Now let a beam of light bounce between the mirrors. Because light travels at the speed limit c, it takes a time $\Delta t' = 2\ell_o/c$ for light to travel from one mirror, call it A, to the second mirror, call it B, and then back to mirror A (see Figure 2.2). Because the clock is at rest in frame S', it is conventional to call the interval $\Delta t'$ "proper time" and denote it $\Delta \tau = 2\ell_o/c$. These simple clocks could be the ones we place along the gridwork in the frame S' to measure the position and time of events.

Now view the clock's operation from the perspective of an observer at rest in S. Now the clock moves to the right at velocity v, and the light ray takes the path in frame S as shown in Figure 2.3. In the frame S, the light ray travels along the line segment $\overline{AB}$ and then along $\overline{BA}$ back to mirror A. The distance mirror A in Figure 2.3 travels between sending and receiving

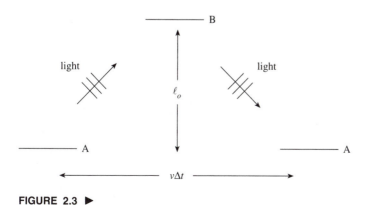

FIGURE 2.3 ▶

the light ray is $v\Delta t$. From Figure 2.3, the distance the light ray travels is

$$\overline{AB} + \overline{BA} = 2\sqrt{\ell_o^2 + (v\Delta t/2)^2}. \tag{2.2.1}$$

But the light ray also travels at the speed limit c in frame S according to Postulate 2, so

$$\overline{AB} + \overline{BA} = c\Delta t. \tag{2.2.2}$$

This is the essential point in this argument—the universal speed limit enters here and we have a clear qualitative distinction from the rules of Newtonian mechanics. Combining Eqs. (2.2.1) and (2.2.2) we find

$$c\Delta t = 2\sqrt{\ell_o^2 + (v\Delta t/2)^2}, \tag{2.2.3}$$

which allows us to solve for Δt,

$$\Delta t = \frac{2\ell_o}{\sqrt{c^2 - v^2}} = \frac{2\ell_o/c}{\sqrt{1 - v^2/c^2}}$$

$$\Delta t = \Delta\tau / \sqrt{1 - v^2/c^2}. \tag{2.2.4}$$

This result is called "time dilation"—the time interval in frame S, Δt, is dilated by a factor of

$$\gamma \equiv 1/\sqrt{1 - v^2/c^2} \tag{2.2.5}$$

compared with the proper time $\Delta\tau$ measured in frame S$'$.

A look at the figure reveals why Δt is greater than $\Delta\tau$, the proper time interval measured on the clock in its rest frame. When the light travels from A to B and back in the frame S, it does so at the speed limit, following Postulate 2. In a Newtonian world where all velocities satisfy the addition of velocities theorem of Galilean transformations, the velocity would be $v_x = v$ and $v_y = c$, producing a speed in frame S of $\sqrt{c^2 + v^2}$. Substituting this into Eq. (2.2.3) leads to the Newtonian result, $\Delta t = 2\ell_o/c = \Delta\tau$ (Newton). But in the real world of Einstein where Eq. (2.2.2) rules, the speed of light is c in all frames, and the time it takes to transverse the path $\overline{AB} + \overline{BA}$ is longer than Newton would calculate. The universal nature of the speed limit produces time dilation directly.

A skeptic might complain that this curious result is particular to the type of clock we made and would not be true of a clock made of springs and masses. We can rule out this objection by appealing to Postulate 1. Consider the clock made of mirrors and light at rest in frame S$'$ and place next to it a mechanical clock made of springs and masses. Synchronize the two clocks. Now do the same in the frame S which is moving at a relative velocity v. This is guaranteed to work in frame S as well as in the original frame S$'$ because of Postulate 1. The two clocks made of mirrors and light experience time dilation, and the mechanical clocks must also because each is synchronized to a clock of mirrors and light at rest with respect to it. The clocks of mirrors and light can be dispensed with at this point and we arrive at the conclusion that the mechanical clocks experience time dilation, too. (Of course, it remains to be seen if we can lay down laws of mechanics that satisfy Postulates 1 and 2. When we discuss the relativistic versions of energy, momentum, and Newton's Second Law, we shall carry out this essential promise.)

As exotic as Eq. (2.2.4) for time dilation might seem, the reader should understand that for relative velocities that are small compared to the speed limit, $c = 3.0 \cdot 10^8$ m/s, the effect is utterly negligible. Take $v = 1000$ mph $\approx$ 490 m/s, so $v/c \approx 1.6 \cdot 10^{-6}$, and using the expansion and approximations derived in Appendix A,

$$\Delta t \approx \Delta \tau \left(1 + \frac{1}{2} \frac{v^2}{c^2} \right) \approx \Delta \tau (1 + 1.33 \cdot 10^{-12}). \qquad (2.2.6)$$

The relativistic correction to the Newtonian result of the equality of Δt and $\Delta \tau$ is suppressed by a factor of the square of the speed limit and is tiny in present everyday situations.

If you live and work at a high energy accelerator center where elementary particles achieve speeds within a fraction of a percent of c, relativistic effects are commonplace. For example, consider the muon, a heavy relative of the electron, which is a common ingredient in cosmic rays and is produced copiously in high energy collisions. It is an unstable particle with a half-life $\Delta \tau \approx 2.2 \cdot 10^{-6}$ s, so a population of muons, at rest in the frame S$'$, decays according to $N(t') = N_o \exp(-0.693 t'/\Delta \tau)$. (The funny factor of 0.693 is just the natural logarithm of 2.) At an accelerator center, this population of muons might be moving in a beam at velocity v comparable to c and the exponential decay law in the frame S, the lab frame, would be $N(t) = N_o \exp(-0.693 t/\gamma \Delta \tau)$. This expression for $N(t)$ has incorporated the quantity $\gamma \Delta \tau$ in its exponential, so the average expected lifetime of a moving muon is dilated to $\gamma \Delta \tau$. Modern accelerators produce beams with $v/c \approx 0.99c$ for which $\gamma \approx 7.1$. This is a huge effect that is easily detected.

For example, if you have a bunch of muons with $N_o = 1000$, they travel across a detector of 1000 m at $v/c = 0.99$ in the time interval $1000/(0.99 \cdot 3 \cdot 10^8) \approx 3.4 \cdot 10^{-6}$ s. Therefore, Newton predicts that $1000 \cdot \exp(-0.693 \cdot 3.4 \cdot 10^{-6}/2.2 \cdot 10^{-6}) \approx 1000 \exp(-1.06) \approx 346$ would survive, because in his world the muon lifetime is $\Delta \tau \approx 2.2 \cdot 10^{-6}$ s in all inertial frames. On the other hand, Einstein predicts that $1000 \exp(-1.06/7.1) \approx 1000 \exp(-0.149) \approx 862$ would survive. The Einstein result (time dilation) is easily confirmed at high energy accelerator centers.

We see that by boosting the muons to a velocity $v/c \approx 1$ we can slow down their internal workings relative to the lab frame. Accelerator centers also make beams of other particles such as protons, which are composed of quarks and gluons, according to modern theories of strongly interacting subnucleon particles. By boosting them to velocities close to the speed of light, we can slow down their internal dynamics so we can probe inside them and learn how quarks interact. We can scatter high energy quanta of light off the accelerated protons and effectively take a snapshot of their internal state when doing high energy scattering experiments. This is how quarks were discovered as the internal particles of protons in the classic experiments at the Stanford Linear Accelerator Center in the 1960s and 1970s.

But there is a puzzle here. If we view the decay process in the frame S' where the muons are at rest, how can we understand that 862 of them survive the trip to the end of the detector and not 346? Obviously, observers in both frames must agree on how many muons get through the detector—both observers can just count them! If relativity is consistent, it must be true that $t'/\Delta\tau = t/\gamma\Delta\tau$, so the two exponential decay laws give identical predictions. We computed $t = d_o/v$, where d_o is the length of the detector in the frame S where it is at rest. Substituting into $t'/\Delta\tau = t/\gamma\Delta\tau$, we learn that $t' = d_o/\gamma v$. How can we understand this? In the frame S' the detector is racing toward the muons at velocity v, so t' should be the length of the *moving* detector divided by the velocity v. It must be that the length of the detector measured in the frame where it is moving is contracted by a factor of γ, $t' = d'/v$ with $d' = d_o/\gamma$! This is indeed the case. The effect is called Lorentz contraction, named after the physicist who first deduced the effect in the context of the theory of electricity and magnetism at the very beginning of the twentieth century. We look more closely at this important effect in the next section.

2.3 Lorentz Contraction

If the constancy of the speed limit implies time dilation, it must also affect the measurements of lengths of moving objects. This effect, called Lorentz contraction, is also easily obtained from Einstein's clock.

To begin, take the clock at rest in frame S' and rotate it by 90° so that its length ℓ_o is parallel to v, the relative velocity of the second frame S. It is conventional to call ℓ_o the proper length of the clock because this is its length in the frame in which it is at rest. Because space is homogeneous in an inertial frame, the rotated clock works as it did before. In frame S', we had $\Delta\tau = 2\ell_o/c$. As before, this proper time interval $\Delta\tau$ must be dilated when measured in frame S because the rotation of the clock can not affect its internal workings. But now view the path of the light ray in the frame S, as shown in Figure 2.4.

We label the length of the clock, from A to B, as measured in frame S as ℓ. In the first image of Figure 2.4, light leaves the mirror and heads toward

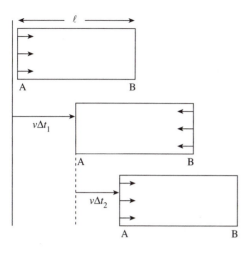

FIGURE 2.4 ▶

mirror B. In the second image, the light ray reaches mirror B after a time Δt_1. The mirror A has moved a distance $v\Delta t_1$ to the right, so

$$c\Delta t_1 = \ell + v\Delta t_1. \tag{2.3.1a}$$

In the third image of Figure 2.4, the light that left mirror B reaches mirror A and is reflected. Between images 2 and 3, a time Δt_2 has passed in frame S, so the mirror A moves an additional distance $v\Delta t_2$,

$$c\Delta t_2 = \ell - v\Delta t_2, \tag{2.3.1b}$$

where the minus sign occurs because the light ray is moving from mirror B to A while mirror A is moving to the right. Finally, the sum of Δt_1 and Δt_2 is the time that passes in frame S corresponding to one period of the clock. We know that this time is Δt and we computed it to be $\gamma(2\ell_o/c)$, where γ denotes, as usual, $1/\sqrt{1 - v^2/c^2}$. So, using Eqs. (2.3.1a) and (2.3.1b),

$$\Delta t = \Delta t_1 + \Delta t_2$$

$$\Delta t = \frac{\ell}{c - v} + \frac{\ell}{c + v} \tag{2.3.2}$$

$$\Delta t = \frac{2\ell/c}{1 - v^2/c^2}.$$

But $\Delta t = \gamma(2\ell_o/c)$, so we learn that

$$\ell = \ell_o/\gamma = \sqrt{1 - v^2/c^2}\,\ell_o. \tag{2.3.3}$$

This is the result we wanted—it is called Lorentz contraction because ℓ is less than ℓ_o. H. A. Lorentz discovered this result in the context of electromagnetism at the beginning of the twentieth century. However, it wasn't until Einstein's derivation that the universality of the result was clarified.

We learn that when an observer measures the length of a rod moving in the direction of its length, he or she will obtain a result contracted by a factor of γ. This derivation shows that this effect follows from the same considerations that led to time dilation—the universal character of the speed limit. But we need to understand this contraction in a more explicit fashion. There are no forces at play here, so what does it mean that the rod is contracted? The effect clearly has nothing to do with how rods are put together—the effect just states how measurements of lengths transform between inertial frames. We see in the next section that Lorentz contraction is only really clarified by a discussion of the synchronization of spatially separated moving clocks.

2.4   The Relativity of Simultaneity

It is clear from our first look at time dilation and Lorentz contraction that the universal nature of the speed of light produces puzzling results. The simplest space-time measurements we can imagine are those that started this chapter—the setting up of a grid of coordinates and clocks—so we will return to it here to understand it better. In particular, let a grid of coordinates and clocks be set up in frame S', which is moving with respect to a frame S at a relative velocity of v to the right. The clocks along the x', y', and z' axes are synchronized in frame S' in the usual way. We want to consider the synchronization procedure from the perspective of observers at rest in frame S. For example, an observer in S' synchronizes two clocks separated by a distance ℓ' in the y' direction by placing a source of light halfway between them, sending a signal out in all directions and insisting that the two clocks read the same time when the signals arrive (Figure 2.5). From the perspective of the observer in frame S, the source and both clocks are moving to the right at velocity v. The light rays both travel at velocity c with respect to frame S, so the light rays arrive at the two clocks simultaneously in frame S. Clearly the synchronization procedure devised in frame S' is

FIGURE 2.5 ▶

acceptable to frame S, and all is well. Also, because each clock is moving with velocity v transverse to the separation between them of ℓ', their positions are $x_1 = vt$, $y_1 = 0$, $z_1 = 0$ for the lower clock and $x_2 = vt$, $y_2 = \ell$, $z_2 = 0$ for the upper clock (no surprises here). These preliminary considerations underlie our discussion of time dilation of the clock with mirrors in Section 2.2.

Things get more interesting when we consider the synchronization procedure for clocks at rest in S' but separated in the x' direction. Let the distance between the clocks in their rest frame S' be $\Delta x' = \ell' = \ell_o$. A signal generator is placed halfway between them, a pulse is sent out in all directions, and the clocks are synchronized. If we say that $t' = 0$ when the light is emitted, then both clocks register $t' = \ell_o/2c$ when the light rays arrive. All observers in all inertial frames will agree that the faces of the two clocks read $t' = \ell_o/2c$ when the light rays are received (Figure 2.6).

Now view the procedure from frame S. An observer in frame S will measure that light reaches clock 1 before it reaches clock 2 because clock 1 is racing toward the rays and clock 2 is racing away from them! The individual times are given by the same sort of considerations we saw in Section 2.3,

$$ct_2 = \ell/2 + vt_2$$
$$ct_1 = \ell/2 - vt_1,$$

(2.4.1)

where t_1 is the time for light to reach clock 1, and t_2 is the time for light to reach clock 2. Note that Eq. (2.4.1) incorporates the universal nature of the speed of light—light, regardless of its direction of motion, travels at speed c with respect to frame S. So we should appreciate, before extracting quantitative details from Eq. (2.4.1), that an observer in frame S concludes that the two clocks are *not* synchronized in his or her rest frame—clocks that are synchronized in one inertial frame are *not* synchronized in another inertial frame if there is a nonzero relative velocity along the line of sight between the clocks. This startling result, which lies at the core of Lorentz

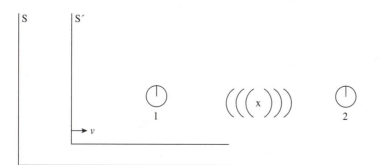

FIGURE 2.6 ▶

contraction, follows simply from the fact that the speed limit is the same in all frames and violates Galileo's law of addition of velocities.

Now let us work out the time difference $t_2 - t_1$ as measured in frame S. The time difference is

$$\Delta t = t_2 - t_1$$

$$\Delta t = \frac{\ell/2}{c - v} - \frac{\ell/2}{c + v} \tag{2.4.2}$$

$$\Delta t = \frac{\ell v/c^2}{1 - v^2/c^2}.$$

But ℓ is the length measured in the frame S moving with velocity v with respect to the frame S′ where the clocks are at rest. So, ℓ is related to ℓ_o, the proper distance between them, by the Lorentz contraction effect $\ell = \ell_o\sqrt{1 - v^2/c^2}$. Thus Eq. (2.4.2) can be written $\Delta t = \gamma \ell_o v/c^2$.

Next we need to calculate the time that an observer in frame S measures as passing in frame S′. Recall from time dilation, that, as measured in frame S, the clocks in frame S′ run slowly. Therefore, if a time $\Delta t'$ passes on a clock at rest in frame S′, then $\Delta t = \gamma \Delta t'$ gives the time interval that passes on a clock at rest in frame S. So, an observer in frame S states that there was a time difference of $\Delta t' = \ell_o v/c^2$ in frame S′ between the time clock 2 received the light ray and the time clock 1 received its ray. But all observers in all frames have noted that clocks 1 and 2 have identical readings when the light rays reach them. Therefore, an observer in frame S concludes that clock 1 was set *ahead* of clock 2 by an amount $\ell_o v/c^2$. In other words, when an observer at time t in frame S inquires what the time t' is in frame S′, he or she finds that t' depends on x'—t' varies as $x'v/c^2$.

The lesson we learn here—that clocks that are synchronized in one frame are *not* synchronized when measured in a frame at relative motion along the line between the clocks—is crucial to understanding time dilation and Lorentz contraction. It is called the "relativity of simultaneity." These three effects must be taken all together (more on this later).

2.5   Time Dilation Revisited

The purpose of this discussion is to understand how time dilation, the relativity of simultaneity, and Lorentz contraction conspire together to produce a consistent picture of time measurements of moving clocks [2]. We have already seen that, if two frames S and S′ are in relative motion, an observer in each says that the other's clock runs slowly. This sounds paradoxical. Let us take a closer look from *both* observers' perspectives and see that each agrees with the observations of the other.

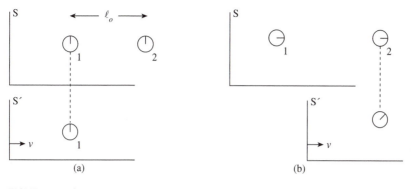

FIGURE 2.7 ▶

Suppose that there are two synchronized clocks at rest in the frame S and separated by a proper distance ℓ_o in the x direction. As usual, let frame S′ be moving to the right at velocity v, as shown in Figure 2.7. Let the clock in frame S′ be synchronized with clock 1, which is at rest in frame S. They pass each other at $x = x' = 0$, $t = t' = 0$, as shown in Figure 2.7a. We wish to know what S′'s clock will read when it passes S's clock 2, as shown in Figure 2.7b. Will observers at rest in both frames make the same prediction even though they see one another's clocks running slowly?

First, consider the measurement from the perspective of an observer at rest in frame S. The clock in frame S′ is moving at velocity v with respect to him or her, so clock 2 will read $t = \ell_o/v$ when the clocks pass. Because the observer in frame S sees frame S′'s clocks running slowly, he or she predicts that S′'s clock will read $t' = \ell_o\sqrt{1 - v^2/c^2}/v$ when the clocks pass each other. So, the observer in S notes that although his or her clock 1 and S′'s clock were synchronized, his or her clock 2 is *not* synchronized with the moving clock!

Now consider the measurements again by treating frame S′ as the one at rest and frame S as moving to the left at velocity v, as shown in Figure 2.8. From the perspective of the observer in frame S′, the distance between clocks 1 and 2 is contracted to $\ell' = \ell_o\sqrt{1 - v^2/c^2}$, so his or her clock will read $t' = \ell_o\sqrt{1 - v^2/c^2}/v$ when it reaches clock 2. So we learn that S and S′ do agree on this point, as they must. Next consider the observation of clock 2 by the observer in S′. This is the tricky part! The left-hand clocks coincide at $t = t' = 0$, as shown in Figure 2.8a. The clock 2 is measured by the observer in S′ to be $\ell_o\sqrt{1 - v^2/c^2}$ to the right and to be $\ell_o v/c^2$ ahead of clock 1! In Figure 2.8b, a time interval $\ell_o\sqrt{1 - v^2/c^2}/v$ has passed in S′, so the observer in S′ states that a time interval $(\ell_o\sqrt{1 - v^2/c^2}/v)\sqrt{1 - v^2/c^2}$ has passed in frame S (this time must dilate to $\ell_o\sqrt{1 - v^2/c^2}/v$ when measured in frame S′). So, clock 2 must read $\ell_o v/c^2 + \ell_o(1 - v^2/c^2)/v$ when the clocks coincide, the observer in S′ reasons. And this agrees with the

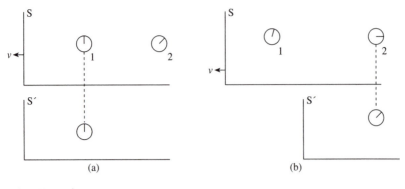

FIGURE 2.8 ▶

observer in frame S, who simply calculates the time by dividing distance by velocity, ℓ_o/v.

In summary, each observer sees the other clocks running slower than his or her own, but, because of the relativity of simultaneity and Lorentz contraction, they both agree that $t = \ell_o/v$ and $t' = \ell_o\sqrt{1 - v^2/c^2}/v$ when the right-hand clock in S coincides with the clock in S'.

2.6 Lorentz Contraction Revisited

Now we want to consider Lorentz contraction again, but this time keep track of the times of the measurements of the positions of the ends of a moving rod [2]. We shall see in this way that the crux of the matter here is the relativity of simultaneity—that spatially separate clocks that are synchronized in one frame are not synchronized in a moving frame. So, when one observer notes the positions of the ends of a rod simultaneously in his or her frame, an observer moving by notes that those measurement events are *not* simultaneous in his frame. This effect, then, explains why observers in relative motion measure different lengths for a given physical rod.

For example, let there be a rod of length ℓ_o at rest in a frame S, which moves to the left at velocity v with respect to frame S' as shown in Figure 2.9. When an observer at rest in frame S' measures the length of the rod, he or she gets an answer ℓ', which must be related to ℓ_o in a linear fashion with, perhaps, a prefactor f that can depend on the dimensionless ratio v/c, $\ell' = f(v/c)\ell_o$. The structure of this formula is a consequence of dimensional analysis. Our task is to find (*and* understand) the prefactor $f(v/c)$. When the observer in S' makes the measurement, he or she notes the positions of the end of the rod at *one instant* in his or her frame, as shown in Figure 2.9, because he or she doesn't want the rod to move during the measurement.

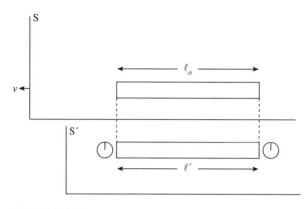

FIGURE 2.9 ▶

    Consider the perspective of the observer in frame S on the measurements. His or her rod of proper length ℓ_o is aligned along the direction of S''s relative velocity v and there is a rod of length ℓ', as measured by the observer in S', moving to the right at velocity v. At $t = t' = 0$, let the left ends of the rods coincide as shown in Figure 2.10a. At this moment the observer in S notes that the clock on the right end of S''s rod reads $t' = -\ell'v/c^2$ by the relativity of simultaneity formula. Later, the right ends of the rods are seen to coincide in S and the clock on that end of the moving rod must read $t' = 0$, as shown in Figure 2.10b, because the observer in S' is making the measurement at one time in his or her frame. Since $t' = -\ell'v/c^2$ is the time on the right-hand clock in Figure 2.10a and is $t' = 0$ in Figure 2.10b, the time elapsed is $\ell'v/c^2$, which is dilated to $\gamma\ell'v/c^2$ in frame S. When the observer in S measures the length of the moving rod, he or she measures a length $f(v/c)\ell' = f^2(v/c)\ell_o$. So, the observer in S can measure the length of the rod ℓ_o at rest in his or her frame by following Figure 2.10: add the length of the moving rod, $f^2(v/c)\ell_o$, to the distance its right end moves,

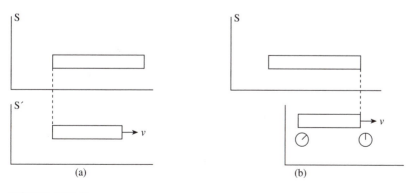

(a)   (b)

FIGURE 2.10 ▶

$(\gamma \ell' v/c^2) \cdot v = \gamma f \ell_o v^2/c^2$. So,

$$f^2 \ell_o + f \gamma \ell_o v^2/c^2 = \ell_o, \tag{2.6.1}$$

which determines f through a quadratic equation,

$$f^2 + \gamma v^2/c^2 f - 1 = 0, \tag{2.6.2}$$

which has solutions

$$f = 1/2(-v^2\gamma/c^2 \pm \sqrt{v^4\gamma^2/c^4 + 4}) \tag{2.6.3}$$

and which can be simplified using $\gamma = 1/\sqrt{1-v^2/c^2}$ to $f = 1/\gamma = \sqrt{1-v^2/c^2}$ or $f = -\gamma$. This last solution is negative and unphysical. So, we have one physical solution,

$$f = \sqrt{1-v^2/c^2}, \tag{2.6.4}$$

and we have successfully rederived the Lorentz contraction formula,

$$\ell' = f(v/c)\ell_o = \ell_o\sqrt{1-v^2/c^2}. \tag{2.6.5}$$

In summary, we see the consistency of the Lorentz contraction formula, $\ell' = \ell_o\sqrt{1-v^2/c^2}$, with time dilation and the relativity of simultaneity. But this time we see how it works and why both observers in relative motion can consistently say that they measure one another's rods as contracted— they observe the ends of the moving rods at different times in the rod's rest frame.

► PROBLEMS ►

2-1. A spaceship (frame S′) passes Earth (frame S) at velocity $0.6c$. The two frames synchronize their clocks at $x = x' = 0$ to read $t = t' = 0$ when they pass one another (call this event 1). Ten minutes later, as measured by Earth clocks, a light pulse is emitted toward the spaceship (call this event 2). Later, the light pulse is detected on the spaceship (call this event 3).

  (a) Is the time interval between events 1 and 2 a proper time interval in the spaceship frame? In the Earth frame?

  (b) Is the time interval between events 2 and 3 a proper time interval in the spaceship frame? In the Earth frame?

  (c) Is the time interval between events 1 and 3 a proper time interval in the spaceship frame? In the Earth frame?

  (d) What is the time of event 2 as measured by the spaceship?

  (e) According to the spaceship, how far away is Earth when the light pulse is emitted?

(f) From your answers in parts (d) and (e), what does the spaceship clock read when the light arrives?

(g) Find the time of event 3 according to Earth's clock by analyzing everything from Earth's perspective.

(h) Are your answers to parts (f) and (g) consistent with your conclusions from parts (a), (b), and (c)?

2-2. Rockets A and B, each having a proper length 100 m, pass each other moving in opposite directions. According to clocks on rocket A, the front end of rocket B takes $1.5 \cdot 10^{-6}$ s to pass the entire length of A.

(a) What is the relative velocity of the rockets?

(b) According to clocks on rocket B, how long does the front end of A take to pass the entire length of rocket B?

(c) According to clocks on rocket B, how much time passes between the time when the front end of A passes the front end of B and the time when the rear end of A passes the front end of B? Does this time interval agree with your answer to part (b)? Should it?

2-3. Short-lived particles are produced in high energy collisions at accelerator centers such as Fermilab in Batavia, Illinois. The study of such particles teaches us about the fundamental building blocks of matter, how nuclei in atoms such as uranium fission and produce energy, how the universe evolved from its origin as a Big Bang, and so on. One particularly important particle that is produced copiously in high energy collisions between protons is the pion. It is the carrier of the nuclear force and has been the subject of much research since the 1950s. Pions decay in their own rest frame according to

$$N(t') = N_o 2^{-t'/T},$$

where T is the half-life, $T \approx 1.8 \cdot 10^{-8}$ s. Imagine that experimenters create pulses of pions at Fermilab and observe that two-thirds of the pions in a particular pulse reach a detector a distance 35 m from the point where they were created. All the pions have the same velocity.

(a) What is the velocity of the pions?

(b) What is the distance from the target to the detector in the rest frame of the pions?

2-4. The distance from Planet X to a nearby star is 12 light-years (a light-year is the distance light travels in 1 year as measured in the rest frame of Planet X). The relative velocity between Planet X and the nearby star is negligible.

(a) How fast must a spaceship travel from Planet X to the star in order to reach the star in 7 years according to a clock fixed on the spaceship?

(b) How long would the trip take according to a clock fixed on Planet X?

(c) What is the distance from Planet X to the nearby star, according to an astronaut on the spaceship?

2-5. A spaceship of proper length ℓ_o travels at a constant velocity v relative to a frame S, as shown in the figure. The nose of the ship (A') passes point A in frame S at time $t = t' = 0$, when a light signal is sent from A' to B', the tail of the spaceship.

   (a) When does the signal reach the tail, B', according to spaceship time t'?

   (b) At what time t_1 in the frame S does the signal reach the tail B'?

   (c) At what time t_2 in the frame S does the tail of the spaceship, B', pass point A?

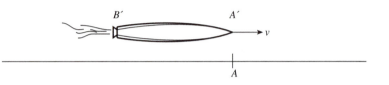

2-6. A flash of light is emitted at position x_1 and is absorbed at position $x_2 = x_1 + \ell$. In a reference frame moving with velocity v along the x axis:

   (a) What is the spatial separation ℓ' between the point of emission and the point of absorption?

   (b) How much time elapses between the emission and absorption of the light?

2-7. At 1:00, a spaceship passes Earth with a velocity $0.8c$. Observers on the ship and on Earth synchronize their clocks at that moment. Just to make this exercise interesting, answer each question from both the viewpoint of Earth and the spaceship.

   (a) At 1:30, as recorded in the spaceship's frame, the ship passes another space probe that is fixed relative to Earth and whose clocks are synchronized with respect to Earth. What time is it according to the space probe's clock?

   (b) How far from Earth is the probe, as measured by Earth's coordinates?

   (c) At 1:30, spaceship time, the ship sends a light signal back to Earth. When does Earth receive the signal by Earth time?

   (d) Earth sends another light signal immediately back to the spaceship. When does the spaceship receive that signal according to spaceship time?

Chapter 3

Visualizing Relativity—
Minkowski Diagrams

3.1 Space and Time Axes for Inertial Frames and the Constancy of Light

The effects discussed in Chapter 2 follow very simply from the two postulates of relativity. The constancy of the speed limit played a central role in them, and, thinking back, we could argue that the qualitative nature of time dilation, Lorentz contraction, and the relativity of simultaneity followed immediately. All this can be done visually using Minkowski diagrams. These pictures of time and position measurements date to the earliest days of relativity. The relativity of simultaneity follows from these diagrams particularly simply and some of our quantitative results can be obtained anew with even less effort.

To begin, set up a space axis x and a time axis t. The transverse directions y and z will be omitted. We choose to display the time t on a clock at rest in frame S as an axis perpendicular to the x axis, as shown in Figure 3.1. Note that we plot ct on the vertical axis, so both axes have units of length. The depiction of light rays is particularly convenient if we do this. It will prove convenient, especially when we discuss the Doppler shift and the Twin Paradox, to use light-years as the units along the ct axis. A light-year is the distance light travels in 1 year. For example, if $ct = 5$ light-years, then t is simply 5 years.

Now consider the synchronization of two clocks at rest in this frame, one at point A and the second at point C. We emit light from a point B

ct

x

FIGURE 3.1 ▶

midway between A and C. The light is recorded on the clocks at events A_1 and C_1, as shown in Figure 3.2.

What do the lines in Figure 3.2 mean? The vertical line $\overline{AA_1}$ indicates that the clock A is at rest in this frame. The fact that the dashed line $\overline{A_1C_1}$ is horizontal indicates that events A_1 and C_1 are simultaneous in this frame. Note that the two light rays, $x = x_B \pm ct$, are drawn at 45° inclined to the x axis—including the factor c in the time axis produces this handy result. In fact, if we shot off a light ray from the origin of the frame to the right, $x = ct$, then its path, called a world line, appears as shown in Figure 3.3.

Minkowski diagrams really become useful when we visualize space-time measurements made in two inertial frames, S and S′, in relative motion. The ct–x axes are laid down. How shall the ct'–x' axes be put on the same figure in a way that is consistent with the postulates of relativity? The ct' axis is the path (world line) in the ct–x graph of the origin $x' = 0$ of the S′ frame. But any point at rest in S′ moves to the right in frame S with velocity v, so $x' = 0$ corresponds to $x = vt$. So, the ct' axis is inclined at an angle θ, $\tan\theta = v/c$, to the ct axis as shown in Figure 3.4. (no surprises, so far).

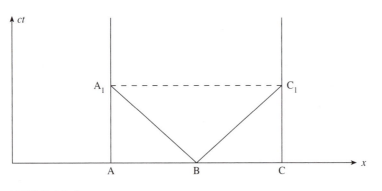

FIGURE 3.2 ▶

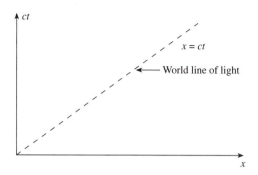

FIGURE 3.3 ▶

But now we need to put the x' axis into Figure 3.4. We have put the world line of a light ray in the figure because the second postulate of relativity, that light travels at the speed limit in any inertial frame, guides us. The light ray bisects the angle between the ct and x axes to guarantee that its velocity is measured as c, $x = ct$. But it must also bisect the angle between ct' and the x' axis, so that its velocity is also c in this frame! So, the x' axis must be tilted at an angle θ, $\tan \theta = v/c$, above the x axis as shown in Figure 3.5.

Because the central results of relativity, time dilation, Lorentz contraction, and the relativity of simultaneity, follow simply from this result, Figure 3.5, let's discuss it in more detail. Note that the $ct'-x'$ coordinate system is not orthogonal. The $ct-x$ and the $ct'-x'$ coordinate systems are not related to one another by a rotation. How do we read off the coordinates of a space-time event in a coordinate system with non-orthogonal axes? Lines of constant ct' are parallel to the x' axis and lines of constant x' are parallel to the ct' axis. So, the event P has the coordinate ct'_p and x'_p, as shown in Figure 3.6.

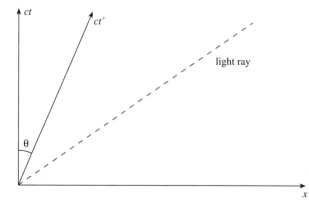

FIGURE 3.4 ▶

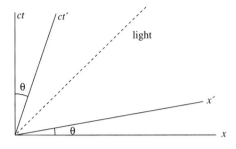

FIGURE 3.5 ▶

Recall that we drew the ct and x axes orthogonally. This was convenient and familiar, but the ct'–x' coordinate system shows that we must learn to use non-orthogonal systems. If the relative velocity is negative, so that S′ is moving to the left in frame S, then Figure 3.6 becomes Figure 3.7. Lines of constant ct', indicating a line of events that are simultaneous in frame S′, are parallel to the x' axis, which has $t' = 0$. The line $\overline{PQ}$ is a line of constant t'. Finally, the line $\overline{PR}$ is a line of constant x' and so is parallel to the t' axis.

Figure 3.5 is so central to relativity that we should derive it in yet another way. The x' axis was drawn so that the frame S′ measures the same speed limit c as does frame S. The x' axis consists of events that are simultaneous in frame S′, events with $t' = 0$. Because this axis is tilted with respect to the x axis, events that are simultaneous in one frame are not simultaneous in the other. We know this from considerations in Chapter 2, but Minkowski diagrams bring that fact to the forefront.

We could also have found the x' axis by boosting the clocks A and C and the signal generator B into the S′ frame, giving them a common velocity v with respect to the S frame. Then the world lines of the clocks would tilt to the right by an angle θ, $\tan \theta = v/c$ (Figure 3.8). We have arranged the ct–x world lines so the signal generator passes $x = 0$ at $t = 0$. The world lines of A, B, and C have a common tilt of θ because they all have a common velocity v with respect to frame S. The dashed line in Figure 3.8 denotes

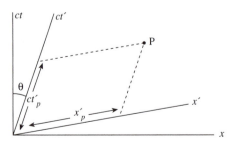

FIGURE 3.6 ▶

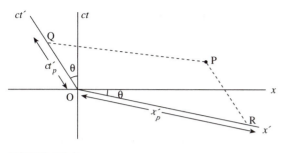

FIGURE 3.7 ▶

the light rays so they are tilted by 45°, one to the right and one to the left, off the t axis. The intersection of the light rays with the world lines of the clocks synchronizes them in the moving frame—the signal generator lies halfway between the clocks and all light rays travel at velocity c with respect to the moving objects, so light reaches the clocks simultaneously in the moving frame. So we have drawn the line $t' =$ constant accordingly in Figure 3.8. It is clearly tilted with respect to the x axis. In fact, the time it takes light to reach clock C is $ct_C = \ell' + vt_C$, and the time it takes light to reach clock A is $ct_A = \ell' - vt_A$. Here ℓ' is the distance, measured in frame S, between the signal generator and one of the clocks. The slope of the t' axis is then

$$\text{Slope} = (ct_C - ct_A)/(ct_C + ct_A). \qquad (3.1.1)$$

Substituting $ct_C = \ell' + vt_C$ and $ct_A = \ell' - vt_A$ into the numerator of Eq. (3.1.1) shows that the slope is v/c. This simply confirms that the x' axis, a line of constant t' along which the moving clocks are synchronized, is tilted at the same angle above the x axis as the ct' axis is tilted below the ct axis. The constancy of the speed limit forces this result on us any way we choose to look at it. Note that ℓ' cancels out of these considerations—

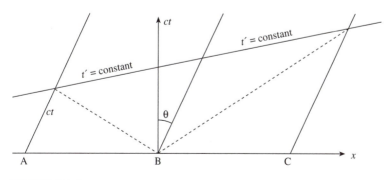

FIGURE 3.8 ▶

all we need to know is that the signal generator lies halfway between the clocks.

3.2 Visualizing the Relativity of Simultaneity, Time Dilation, and Lorentz Contraction

Let's use Minkowski diagrams to understand the three basic relativistic phenomena we have been discussing. The Minkowski diagram, because it shows both the ct–x and ct'–x' axes, will allow us to view all the phenomena from both points of view and see how paradoxes are avoided.

First consider the relativity of simultaneity. In Figure 3.9 we show two clocks at rest in frame S and separated by a proper distance ℓ_o. We show the two clocks at $t = 0$ and at some later time t. We also show the x' axis, points of constant t', for the frame S' moving with velocity v, $\tan \theta = v/c$. We see instantly that the clocks at rest in frame S are not synchronized in frame S' if they are separated in the direction x. We also see that clock 1 is behind clock 2 when viewed in frame S'. This confirms the result of Chapter 2, but in a totally transparent fashion. In fact, it is trivial to calculate the basic formula of the relativity of simultaneity from the geometry of Figure 3.9. We see in the figure that the distance $d = \ell_o \tan \theta = \ell_o v/c$. Because $d = ct$, we have $t = \ell_o v/c^2$, the time difference between clocks 1 and 2 in Figure 3.9 at a specified t'.

Even more interesting, we can now turn the tables and view clocks at rest in frame S' from the perspective of an observer at rest in frame S. Consider the world line of two clocks at rest in S' in Figure 3.10.

The clocks on the x' axis ($t' = 0$) are synchronized as shown. Therefore, when compared at the same value of t, along the x axis as shown in the figure, clock 2 is behind clock 1. The moving clock that is spatially ahead

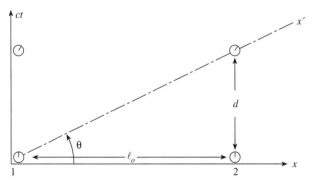

FIGURE 3.9 ▶

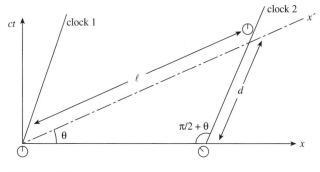

FIGURE 3.10 ▶

lags in time. The quantitative effect can also be read off Figure 3.10. By the law of sines, we read off the triangle,

$$\frac{d}{\ell} = \frac{\sin\theta}{\sin(\pi/2+\theta)} = \frac{\sin\theta}{\cos\theta} = \tan\theta = \frac{v}{c}. \tag{3.2.1}$$

Because $d = ct'$, the time difference between the clocks in frame S′, we again have $t' = \ell v/c^2$, where ℓ is the distance between the clocks in their rest frame S′.

Now let us turn to the visualization of Lorentz contraction using Minkowski diagrams. First, however, we must settle on the units of length and times used on the unprimed and primed axes in the diagram. The relation of space-time measurements in frame S and frame S′ does not preserve angles, as is evident from our Minkowski diagram pictures. It also does not preserve lengths as we have seen in our discussions of time dilation and Lorentz contraction. This means that the scale of lengths of the x' and ct' axes relative to the x and ct axes is not unity. To see this, consider two frames in relative velocity v and $\gamma = 1/\sqrt{1 - v^2/c^2}$. Consider a rod of unit length at rest in the moving frame S′. By Lorentz contraction, we measure this length if we observe a rod of length γ in the frame S. We show this situation in Figure 3.11—the rod at rest in frame S lies along $\overline{OA}$, which has a length γ. The world lines of the ends of the rod are shown as the vertical lines in the Minkowski diagram.

Because $\overline{OA} = \gamma$ units of length in the S frame and because $\tan\theta = v/c$, the segment $\overline{AB} = v\gamma/c$. Therefore, the length $\overline{OB}$, which is the length of the rod as measured in the frame S′, is

$$\overline{OB} = \sqrt{\frac{v^2}{c^2}\gamma^2 + \gamma^2} = \sqrt{\left(1 + \frac{v^2}{c^2}\right)\gamma^2} = \sqrt{\left(1 + \frac{v^2}{c^2}\right)\bigg/\left(1 - \frac{v^2}{c^2}\right)}. \tag{3.2.2}$$

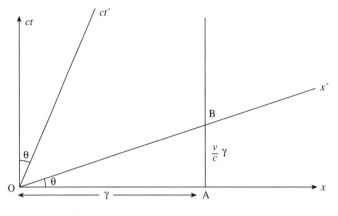

FIGURE 3.11 ▶

But $\overline{\text{OB}}$ represents the length of a rod of unit length in the S′ frame. So, in the Minkowski diagram there is a change of scale between the frames:

$$\text{S}'/\text{S} = \sqrt{\left(1+\frac{v^2}{c^2}\right)\Big/\left(1-\frac{v^2}{c^2}\right)}.\qquad(3.2.3)$$

This effect only occurs when we visualize lengths of two frames on a single Minkowski diagram. It will show up in no other parts of our analysis, and we need it just to draw our figures accurately.

To visualize Lorentz contraction, choose $v/c = 3/5$, so the arithmetic works out neatly and $\gamma = 1/\sqrt{1-v^2/c^2}$ is exactly $\gamma = 5/4$. The scale factor is $\text{S}'/\text{S} = \sqrt{34}/4 \approx 1.46$. So, if we consider a rod of unit length at rest in the frame S′, we generate the Minkowski diagram in Figure 3.12, where the length of the rod of unit proper length is measured as having a length $\Delta x = 1/\gamma = 1/(5/4) = 0.8$ in the frame S, by the Lorentz contraction formula. Turning the tables, a rod of unit length at rest in the frame S, whose ends sweep out the world lines $x = 0$ and $x = 1.0$, the dashed line, is measured to have length 0.8 along the x' axis. Again, we see from Figure 3.12 that each observer measures the length of a moving rod as contracted by a factor of 1.25. The reason for this contraction is the fact that simultaneous measurements in one frame are not simultaneous in a moving frame.

The visualization of time dilation is similar. Consider a clock at rest in frame S and let frame S′ be moving to the right at relative velocity $v/c = 3/5$ so $\gamma = 5/4$ again. The t and t' axes are shown in Figure 3.13. We have drawn the tick marks on the ct' axis to scale, $\text{S}'/\text{S} = \sqrt{34}/4 \approx 1.46$. At $ct = 5$ the observer at rest in frame S notes that the clock at rest in frame S′ reads $ct' = 4$ because the dashed horizontal line of $ct = 5$ intercepts the

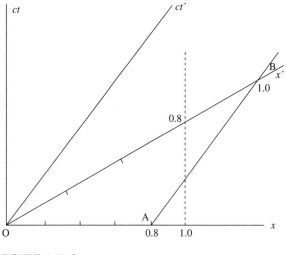

FIGURE 3.12 ▶

ct' axis at $ct' = 4$. This is time dilation, $ct = \gamma ct'$, which reads $5 = (5/4) \cdot 4$. The observer at rest in frame S states that the moving clock at rest in frame S′ is running slowly by a factor of $1/\gamma$. Similarly, the observer at rest in frame S′ notes that where his clock reads $ct' = 5$ the clock in frame S reads $ct = 4$ (follow the dash-dot line in Figure 3.13).

In summary, both observers agree that moving clocks run slowly. There is no contradiction in this statement because the two observers do not share the same time axis.

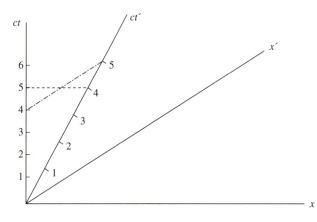

FIGURE 3.13 ▶

3.3 | The Doppler Effect

Everyone knows that a train whistle sounds higher when it is approaching and lower when it is receding. A similar effect occurs in relativity when we observe a moving wave. We need the quantitative details of this effect for light before we analyze the Twin Paradox. The Doppler effect is very important in astronomy, so it is worth examining in some detail.

Suppose that there is a signal generator at rest in frame S', which is moving at velocity v with respect to frame S, and an observer at rest in frame S sees the signal generator approaching him or her (Figure 3.14). Let the wave train consist of n cycles that the observer detects in the course of a time interval Δt. Because light moves at velocity c with respect to frame S and the signal generator approaches at velocity v, the length of the wave train, $n\lambda$, observed in frame S, is

$$n\lambda = c\Delta t - v\Delta t, \tag{3.3.1}$$

where λ is the wavelength of the light in frame S. Since wavelength and frequency are related by $\nu\lambda = c$, we can write Eq. (3.3.1) as

$$n = \nu(1 - v/c)\Delta t. \tag{3.3.2}$$

But we can also write n in terms of quantities observed in frame S'. The source is at rest there and produces a frequency ν_o, say, over a time interval $\Delta t'$, so another expression for n is

$$n = \nu_o \Delta t'. \tag{3.3.3}$$

But $\Delta t'$ is the proper time interval for the signal generator, so Δt is given by time dilation,

$$\Delta t = \gamma\Delta t', \tag{3.3.4}$$

and Eqs. (3.3.2)–(3.3.4) give

$$\nu_o\Delta t/\gamma = \nu(1 - v/c)\Delta t. \tag{3.3.5}$$

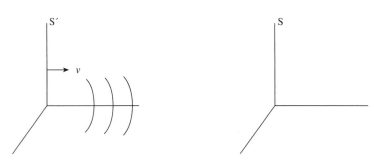

FIGURE 3.14 ▶

Solving for the frequency detected in frame S,

$$\nu = \nu_o/\gamma(1 - v/c) = \sqrt{\frac{1 + \frac{v}{c}}{1 - \frac{v}{c}}} \nu_o \qquad \text{(approaching)}. \qquad (3.3.6)$$

So, the observer at rest in frame S measures the frequency of the light to be higher due to the approaching motion of the source.

Clearly, if the source had been receding we would have $v \rightleftharpoons -v$,

$$\nu = \sqrt{\frac{1 - \frac{v}{c}}{1 + \frac{v}{c}}} \nu_o \qquad \text{(receding)}. \qquad (3.3.7)$$

This effect is particularly important in astronomy where it produces the red shift of distant receding stars—the frequency of light coming from rapidly receding stars is reduced, as described by Eq. (3.3.7).

Consider a numerical example we use later in a discussion of the Twin Paradox. Let a source generate a light pulse once a year and suppose it moves away from us at $v/c = 0.8$. How long is the interval between the pulses that we receive? Substituting into Eq. (3.3.7), we find

$$\nu = \sqrt{\frac{1 - 0.8}{1 + 0.8}} \nu_o = \nu_o/\sqrt{9} = \nu_o/3. \qquad (3.3.8)$$

So, we receive one signal every 3 years. If the source were moving toward us at $v/c = 0.8$, then we would observe $\nu = 3\nu_o$; that is, we would receive a signal every 4 months.

It will prove instructive to visualize the Doppler shift effect on a Minkowski diagram. Let the source be at rest in frame S and let the frame S′ recede at $v/c = \tan \theta = 0.8$. The Minkowski diagram is shown in Figure 3.15.

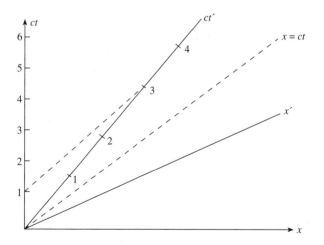

FIGURE 3.15 ▶

We have set the scale using $S'/S = \sqrt{(1 + v^2/c^2)/(1 - v^2/c^2)} = \sqrt{41}/3 \approx$ 2.13. We see in Figure 3.15 that the clock at rest in frame S emits a light pulse at $ct = 1$, which is detected by the clock at rest in frame S' at $ct' = 3$. We were careful in the figure to set the relative scales of the time axes when we drew the tick marks and we were careful to draw the light ray with slope 1. We will see that the Doppler shift gives us a convenient way of monitoring the readings on clocks in relative motion. This will be handy in our discussion of the Twin Paradox.

3.4 The Twin Paradox

Probably the most famous paradox in all of science is the Twin Paradox. We will resolve this paradox in several ways because it overthrows the notion of Newtonian time so dramatically. We do so in the context of Minkowski diagrams, although the algebra that could accompany the diagrams would do as well.

The paradox is the following. Suppose there are two identical twins, Mary and Maria. They have a terrible fight and Maria rockets away at a velocity of 80% of the speed of light, $\tan \theta = v/c = 4/5$ for 5 years, as measured by Mary, at rest in frame S. Then Maria has a change of heart. She returns as quickly as she departed and throws herself into Mary's arms as Mary's clock strikes a decade. They embrace and make up, but something is not quite right—Maria looks considerably younger than Mary. In fact, Maria's clock indicates that only 6 years has passed! Yes, indeed, Maria is now 4 years younger than Mary. They are no longer identical twins!

Is this possible? Mary, who knows relativity, tries to understand the result by noting that Maria had a relative velocity of $v/c = 0.8$ throughout a decade, so 10 years must be the time-dilated interval of the time passed on a wristwatch attached to Maria. Because $\gamma = 1/\sqrt{1 - 0.8^2} = 5/3$, time dilation implies $\Delta t = \gamma \Delta t'$ or $10 = (5/3) \cdot 6$, Mary feels that she understands the time interval involved. But Maria, distraught at now being Mary's "younger" sister, contends that Mary was, from her perspective, racing away at a relative velocity $0.8c$ and she should have aged more slowly! How can both see one another's clocks as running slowly and a paradoxical situation not be found when their trips are over?

In light of our earlier discussions of time dilation, Lorentz contraction, and the relativity of simultaneity, the reader, hopefully, is not as puzzled by this apparent paradox as the uninitiated. Let's go through the Minkowski diagram analysis of the trips and see that all works out. Yes, Maria ends up 4 years younger than Mary and has to live with that—at least until the next trip.

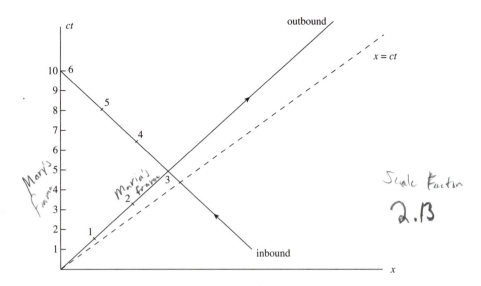

FIGURE 3.16 ▶

The trip is shown in Figure 3.16. Mary and her clock reside at $x = 0$ and 10 years click off. Maria dashes off at $v/c = 0.8$ for 5 years according to Mary and then returns at $v/c = -0.8$ for another 5 years. From Mary's perspective, Maria's clock, her wrist-watch, say, is running slowly and $\Delta t = \gamma \Delta t'$ applies with $\gamma = 1/\sqrt{1 - v^2/c^2} = 1/\sqrt{1 - 0.8^2} = 5/3$, so Maria ages 3 years during her outbound trip and 3 more years during her inbound trip.

All of this seems clear, except the portion of Maria's trip where she turns around. At this point she experiences acceleration and her frame is no longer inertial. There is a simple way to deal with this. We could consider the outbound leg of the trip as made by one rocket with $v/c = 0.8$ and the inbound trip as made by *another* rocket with $v/c = -0.8$. Let the inbound and outbound rockets pass closely by one another at $ct = 5$ and synchronize their clocks. In this way we never have to consider acceleration and our analysis holds without compromise.

To get the full understanding we seek, we must consider the trip from the perspective of Maria. She travels outbound for 3 years as measured on her wristwatch. From Maria's perspective, Mary is racing away at $v/c = -0.8$. Maria observes Mary's clock running slowly and records that Mary ages by an amount $\Delta t'/\gamma = 3/(5/3) = 9/5 = 1.8$ years. We see this in the Minkowski diagram, as shown more explicitly in Figure 3.17, where the line $\overline{AB}$ of constant t' is parallel to the x' axis. So, Maria measures that Mary has aged only 1.8 years throughout her outbound trip.

Now, something special happens at the point B where Maria turns back toward her sister—the line of constant t' tilts because we are jumping onto a new inertial frame, one with $v/c = -0.8$. This frame is shown in Figure 3.18.

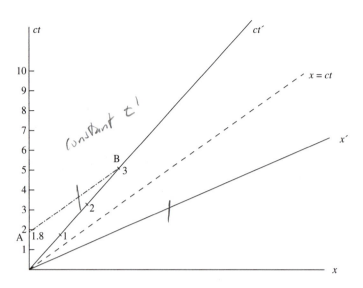

FIGURE 3.17 ▶

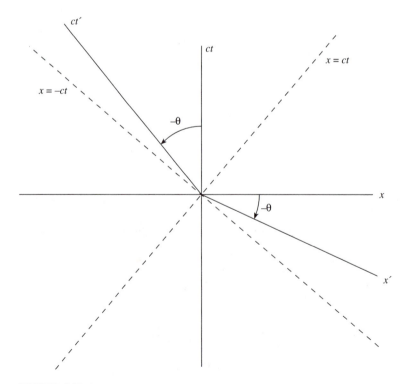

FIGURE 3.18 ▶

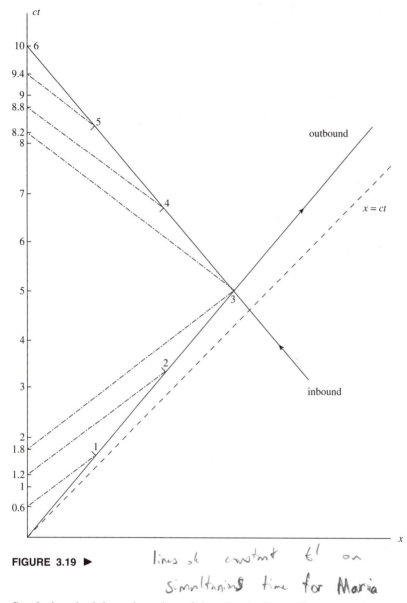

FIGURE 3.19 ▶   lines of constant t' on
simultanins time for Maria

So, during the inbound portion of the trip, the lines of constant t' are parallel
to the x' axis in Figure 3.18, which tilt by angle $-\theta$ *below* the x axis.

The lines of constant t' are shown in Figure 3.19 for both legs of the
trip. We see that the turnaround, where $ct' = 3$ light-years, corresponds
to both $ct = 1.8$ light-years *and* $ct = 8.2$ light-years. Then, when $ct' =$
4, $ct = 8.8$ light-years, and when $ct' = 5, ct = 9.4$ light-years. It is quite
interesting that Maria concludes that $8.2 - 1.8 = 6.4$ years pass on Mary's

clock during the turnaround! Maria can now understand how 10 years pass on Mary's clock, even though Maria measures that Mary's clock runs slower than her own!

Another perspective on this curious sequence of events is afforded by the Doppler effect. This was first illustrated by the mathematician C. G. Darwin [6]. Mary and Maria can monitor one another's clocks by agreeing to send one another signals on their birthdays. So, Mary sends a greeting to Maria on each of her 10 birthdays. This is shown in Figure 3.20.

Mary sends a birthday greeting at $ct = 1$ light-year, and Maria receives it at the turnaround point $ct' = 3$ light-years, as shown. This agrees with the Doppler formula $\nu = \sqrt{(1 - v/c)/(1 + v/c)} = \sqrt{(1 - 4/5)/(1 + 4/5)} = 1/3$. So, 3 years pass in Maria's frame before she receives the birthday greeting from Mary.

Maria's positon is $x_m = (4/5)ct$ during the outbound trip, and the light ray sent by Mary on her first birthday travels on $x = c(t - 1)$. So, Maria receives the signal at $(4/5)t = t - 1$ or $t = 5$. Maria is now 4 light-years from Earth and is about to turn around. The next eight light rays sent by Mary and received by Maria are shown in the figure. Maria receives nine signals during the second half of her journey (the inbound part) and just one during the first half. She receives 10 signals and knows that Mary ages 10 years while she ages only 6 years. The Doppler formula for the inbound part of Maria's journey reads $\nu = \sqrt{(1 + v/c)/(1 - v/c)} = \sqrt{(1 + 4/5)/(1 - 4/5)} = 3$, so Maria receives three signals each year from Mary for the duration of

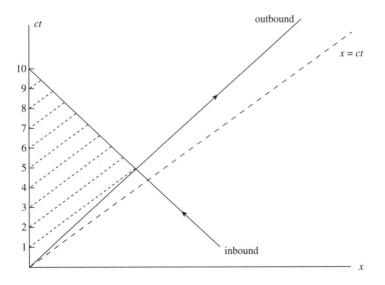

FIGURE 3.20 ▶

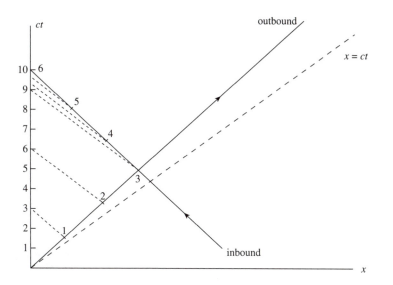

FIGURE 3.21 ▶

the inbound journey, 3 years on Maria's wristwatch, accounting for nine signals.

Finally, we turn the tables again and suppose that Maria sends greetings to Mary on each of her birthdays. This situation is shown in Figure 3.21. During the outbound trip, Maria sends three greetings and Mary receives one every 3 years, at years 3, 6, and 9 on her clock. Then after the turnaround, Maria sends three more greetings and Mary receives them every 4 months between years 9 and 10 on her wristwatch. Again, both Mary and Maria agree that Mary has aged 10 years while Maria has aged 6.

3.5 Einstein Meets Shakespeare—Relativistic History

The Twin Paradox illustrates how much of present culture and history is tied to the Newtonian concept of absolute time. In a world where space travel at relative speeds approaching the speed limit would be possible, everything would be so different. For example, the ascendancy of the throne in the days of the kings of England would have to be revised. The very idea that the son of the king of England will later become the king who will have another son who will then become king, and so on, is rooted in the notion of absolute time. For example, say that Prince William was jealous of his older brother Prince Harry because Prince William has little chance to rule as king. Well, he could always take a relativistic journey akin to that taken by Maria and time things so that on his return King Harry, who, we suppose, is childless, is on his deathbed while he, Prince William, has aged only a few years. Shortly, then, Prince William would become King William and rule a long and glorious time—barring other surprises!

▶ Problems ▶ _____

3-1. Two inertial coordinate systems S and S′ are moving with respect to one another at a relative velocity $c/2$. Draw a Minkowski diagram relating these two frames.

   (a) Calibrate the axes and mark a unit length along each following Eq. (3.2.3).

   (b) Plot the events on the diagram: (i) $ct = 1$, $x = 1$; (ii) $ct = 2$, $x = 0$; (iii) $ct′ = 1$, $x′ = 1$; (iv) $ct′ = 2$, $x′ = 0$.

   (c) Take each event in part (b) and determine from the Minkowski diagram its coordinates in the other frame.

3-2. Radar Trap! A policeman aims his stationary radar transmitter backward along the highway toward oncoming traffic. His radar detector picks up the reflected waves and analyzes their frequencies. Suppose that his transmitter generates waves at a frequency ν_o and detects the waves reflected by an approaching speeding car at frequency ν_r.

   (a) Draw a Minkowski diagram showing the world lines of the stationary policeman, the approaching speeding car, and several transmitted and reflected radar waves.

   (b) From the geometry of your Minkowski diagram relate the time between transmitted radar waves to the time between reflected waves, as detected by the stationary policeman and derive, $\nu_r = [(1 + v/c)/(1 - v/c)]\nu_o$.

   (c) Use this result to derive the Doppler shift formula obtained in the chapter. Call the frequency that an observer in the speeding car measures $\nu′$. Argue on the basis of the first principle of relativity that $\nu′/\nu_o = \nu_r/\nu′$. Combine this result with part (b) to derive the standard Dopper shift formula, $\nu′ = \sqrt{(1 + v/c)/(1 - v/c)}\,\nu_o$.

   (d) Suppose that the car's speed is 100 mph and $\nu_o = 10^{10}$ cps. Predict ν_r approximately by linearizing the formula in part (b). (Because 100 mph is tiny compared to the speed of light, this approximate treatment is very good.) Can handheld radar equipment carried by your local law enforcement department detect such a small fractional frequency shift? [Ref: T. M. Kalotas and A. R. Lee, *Am. J. Physics* 58, 187 (1990).]

3-3. Transverse (Quadratic) Doppler Shift. The Doppler shift we have discussed so far occurs when the transmitter and the detector of the waves are heading directly toward or away from one another. It might happen, however, that the transmitter and the detector approach one another at an angle θ as shown in the figure. The figure shows the transmitter with a velocity v traveling past an observer with the distance of closest approach, labeled R. When $\theta = 90°$, we have the transverse Doppler shift.

(a) Show that there is a Doppler shift in the case $\theta = 90°$, which is given just by time dilation. Calling ν_o the frequency of light in the rest frame of the transmitter and ν' the frequency of light in the rest frame of the observer, at the origin in the figure, derive $\nu' = \nu_o/\gamma$, where $\gamma = 1/\sqrt{(1 - v^2/c^2)}$, as usual. Note that for small v^2/c^2, $\nu' \approx (1 - v^2/2c^2)\nu_o$, and hence the name quadratic Doppler shift.

(b) Armed with the result of part (a) and the standard Doppler shift formula for head-on motion, obtain the Doppler shift formula for the general case shown in the figure, $\nu' = \nu_o/[\gamma(1 - (v/c)\cos\theta)]$. (*Hint*: Resolve the relative motion between the transmitter and the observer into radial motion, along the radius r shown in the figure, and transverse motion.)

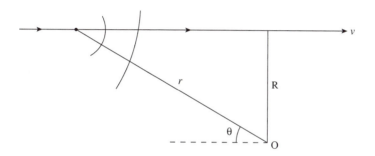

3-4. When we resolved the Twin Paradox, we noted that from Maria's perspective, Mary ages 6.4 years as Maria jumps from the outgoing to the incoming rocket. Obtain this numerical result from our relativity of simultaneity formula, vx/c^2, by noting that v changes by $2v$ as Maria jumps between the rockets. Choose $v = 0.8c$.

3-5. Here is a relatively easy paradox based on an incomplete understanding of time dilation—your task is to resolve it.

Suppose that frame S', a huge transport rocket, moves in the x direction relative to frame S, a space station, at velocity v, as usual. Suppose that there is an astronaut in the a huge transport rocket, who has a rocket-powered backpack and is traveling in the $-x$ direction at speed $-v$ relative to the transport rocket. Therefore, our astronaut is actually at rest with respect to the space station. A critic of relativity might argue that time dilation states that clocks at rest in the transport rocket run slowly compared to those at rest in the space station, and the astronaut's clock runs slowly compared to those at rest in the transport rocket. Therefore, the astronaut's clock is predicted to run slow-squared compared to those at rest in the space station! But the astronaut is actually at rest in the space station, so we have an apparent contradiction!

What's wrong with this argument? Clarify it with a Minkowski diagram showing the three relevant time axes and spatial axes of the three frames involved.

3-6. Consider Problem 2.1 once more. Diagram the events and world lines of that problem in a Minkowski diagram. Establish the scales on the time and spatial axes so you can make quantitative predictions from the Minkowski diagram itself in either frame. Answer Problem 2.1(a)–(h) directly from your Minkowski diagram.

Chapter 4

Assorted Applications

4.1 Lorentz Transformation

Now that we understand how relativity works, let's derive the generalization of the Galilean transformation relating the time and position of an event measured in one inertial frame to those measurements in another inertial frame. Recall the Galilean result. We have a frame S′ moving to the right in frame S at relative velocity v. Then if a point x' is observed at time t' in frame S′, an observer in frame S would assign the values

$$x = x' + vt \tag{4.1a}$$

$$t = t' \tag{4.1b}$$

to the point, assuming that the origin of the two frames, $x = x' = 0$, coincided at $t = t' = 0$.

In relativity, Eqs. (4.1a) and (4.1b) are replaced by the Minkowski diagram shown in Figure 4.1. Instead of the picture, we want explicit formulas for x and t, given x' and t' for an event in Figure 4.1. Armed with our knowledge of Lorentz contraction, time dilation, and the relativity of simultaneity, this is a snap.

Consider the relativistic generalization of Eq. (4.1a) in Figure 4.2. Galileo states that the distance vt and x' add up to x. However, because x' is measured in frame S′, Lorentz contraction shrinks it to x'/γ when observed in frame S. So, Eq. (4.1a) is replaced by

$$x = x'/\gamma + vt,$$

which can be solved for x',

$$x' = \gamma(x - vt). \tag{4.2a}$$

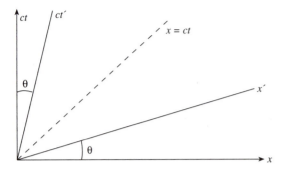

FIGURE 4.1 ▶

This tells us how the x' axis in the Minkowski diagram Figure 4.1 is related to the x and ct axes there.

Next we need the generalization of Eq. (4.1b). In other words, we need to know how the ct' axis in the Minkowski diagram Figure 4.1 is related to the ct and x axes there. But this relation must have the same form as Eq. (4.2a), in order that both frames S and S′ observe the same speed of light. In other words, take Eq. (4.2a) and replace x' with ct', x with ct and t with x/c. So,

$$ct' = \gamma(ct - vx/c)$$

or

$$t' = \gamma(t - vx/c^2). \tag{4.2b}$$

Once more with emphasis—the central idea in this derivation is the constancy of the speed of light—once we know how x' is related to x and t, t' must be related to x and t so that both frames measure the identical speed of light, as expressed pictorially in the Minkowski diagram.

Eqs. (4.2a) and (4.2b) constitute the Lorentz transformations. It is a central result of special relativity. It incorporates all we know about time dilation, Lorentz contraction, and the relativity of simultaneity. In order to

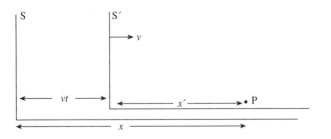

FIGURE 4.2 ▶

see how the formulas work, let's obtain our three basic results by taking special cases of Eqs. (4.2a) and (4.2b).

4.1.1 Time Dilation

Consider two times t_2 and t_1 that occur on a clock at rest in frame S. From Eq. (4.2b) an observer in S' would measure the time interval

$$t_2' - t_1' = \gamma \left[t_2 - t_1 - \frac{v}{c^2}(x_2 - x_1) \right] = \gamma(t_2 - t_1),$$

where we used the fact that the clock is at rest in frame S, so $x_2 = x_1$. This is our usual time dilation formula because $t_2 - t_1 = \Delta\tau$, the proper time interval.

4.1.2 Lorentz Contraction

Consider a rod at rest in frame S' extending from x_1' to x_2'. Its length is measured in frame S by noting the corresponding x_1 and x_2 at a given time $t_2 = t_1$. Eq. (4.2a) predicts

$$x_2' - x_1' = \gamma(x_2 - x_1).$$

Because $x_2' - x_1' = \ell_o$ is the proper length of the rod, we have Lorentz contraction, $\Delta x = \ell_o/\gamma$.

4.1.3 Relativity of Simultaneity

Consider five clocks at rest and synchronized in frame S [1]. Let each clock be separated from its neighbor by ℓ_o (Figure 4.3). Now measure the time of each clock in frame S', as shown in the Minkowski diagram (Figure 4.4). We observe the times on each clock at a given single time t' in frame S', as shown in the diagram. Let us apply Eq. (4.2b) and note that clock C at $x = 0$ gives the value of t' directly, $t' = \gamma t_c$. So,

$$\gamma t_c = \gamma \left(t - \frac{v}{c^2} x \right),$$

FIGURE 4.3 ▶

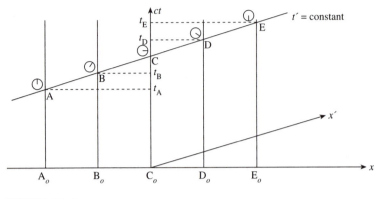

FIGURE 4.4 ▶

which we solve for t,

$$t(x) = t_c + \frac{v}{c^2}x \qquad t' = \text{constant}. \tag{4.3}$$

This result expresses our relativity of simultaneity effect—the clocks that are moving in frame S are not synchronized, with the leading clock A behind the trailing clock E by an amount $v\Delta x/c^2$.

The Lorentz transformation formulas are quite convenient to apply, but they frequently fail to teach us as much as more elementary but challenging arguments.

4.2 Relativistic Velocity Addition

One of the hallmarks of Newtonian physics is the "obvious" fact that velocities add. Consider the Galilean transformation Eq. (4.1a) and let the particle move at velocity v'_p relative to frame S'. What then is its velocity, $v_p = x/t$, with respect to frame S? Substituting into Eq. (4.1a) we have

$$v_p t = v'_p t' + vt.$$

Using the absolute nature of Newtonian time, $t = t'$, we see that

$$v_p = v'_p + v. \tag{4.4}$$

This result is used in day-to-day physics problems and serves us well as long as all the velocities are small compared to the speed limit c. It is clearly inconsistent with special relativity because it doesn't respect the existence of a speed limit—by adding velocities linearly we can generate a velocity for a physical particle as high as we like.

We need to know what replaces Eq. (4.4) in special relativitiy. To answer this, we go back to the Lorentz transformations, Eqs. (4.2a) and (4.2b) and substitute in $x' = v_p' t'$ and $x = v_p t$ to describe a particle in uniform motion. Now Eq. (4.2a) becomes

$$v_p' t' = \gamma(v_p t - vt) = \gamma(v_p - v)t. \tag{4.5a}$$

Eq. (4.2b) becomes

$$t' = \gamma\left(t - \frac{v}{c^2} v_p t\right) = \gamma\left(1 - \frac{vv_p}{c^2}\right)t. \tag{4.5b}$$

Substituting back into Eq. (4.5a), we have

$$v_p' = \frac{v_p - v}{1 - vv_p/c^2}. \tag{4.5c}$$

Or we could solve for v_p,

$$v_p = \frac{v_p' + v}{1 + vv_p'/c^2}. \tag{4.5d}$$

Note that this result is nonlinear in v and v_p' in just the right way to enforce the speed limit. For example, let the particle move at the speed limit in frame S′, $v_p' = c$. Then,

$$v_p = \frac{c + v}{1 + vc/c^2} = c \cdot \frac{c + v}{c + v} = c.$$

In other words, light can have velocity c with respect to both frames S and S′ even if they are in relative motion! This result was built into our formalism by the second postulate of relativity, but it is still interesting to see how it is enforced. The reader should check the even more peculiar case of letting $v_p' = -c$ in Eq. (4.5d). Again we find $v_p = -c$ for *all* v, no matter how large!

This derivation also suggests that although transverse distances y and z are uneffected by the boost between frames S and S′,

$$y' = y$$

$$z' = z,$$

it is not true that transverse velocities are the same in both frames. This is so because of time dilation. Let $y' = u_p' t'$, motion just in the y' direction in the frame S′. This particle will have a velocity in the x direction in frame S of v and a y component of velocity $y = u_p t$. Now,

$$u_p' = \frac{y'}{t'} = \frac{y}{t'} = \frac{u_p t}{t'} = \frac{u_p}{\gamma(1 - vv_p/c^2)}. \tag{4.6}$$

This ugly, complicated formula involves both the velocities in the x direction, v and v_p, and those in the y direction, u_p and u'_p. We will need Eq. (4.6) later when we discuss dynamics and relativistic momentum.

It is interesting to revisit Eq. (4.5d) and ask for the source of the crucial nonlinearity, the denominator, $1 + vv_p/c^2$. Which of our three effects, Lorentz contraction, time dilation, or the relativity of simultaneity, is responsible here? The γ factors in Eqs. (4.5a) and (4.5b) cancel out of (4.5c), so Lorentz contraction and time dilation are not the culprits here. Inspecting Eq. (4.5b), we see the relativity of simultaneity at work—the fact that a moving clock that is attached to the particle is increasing its x' coordinate as $v'_p t'$, produces the nonlinearity that makes the speed of light truly universal.

4.3 | Causality, Light Cones, and Proper Time

When observers measure physical phenomena, such as the position and time of events, they record space-time coordinates whose values depend on their coordinate systems. Some quantities are the same in all reference frames and they are particularly useful and significant. The speed of light is such an invariant—it is measured to be the speed limit, $3.0 \cdot 10^{10}$ cm/s, in all inertial frames. The proper times of clocks and proper lengths of rods are also invariants. This last remark might sound trivial—to measure the proper time of a clock we must boost ourselves to the rest frame of the clock of interest and note a proper time interval. Similarly for a proper length of a rod. The term "proper" indicates exactly how the measurement is to be done.

All this would be more interesting if we could infer the proper time interval passing on a clock by making measurements in another inertial frame. Suppose the clock is at rest in frame S and a proper time $\Delta\tau$ passes. In another inertial frame S', a time interval $\Delta t'$ will pass while the clock moves a distance $\Delta x' = -v\Delta t'$. $\Delta t'$ is related to $\Delta\tau$ by time dilation: $\Delta t' = \Delta\tau/\sqrt{1 - v^2/c^2}$. Now we need a relation between $\Delta\tau$, $\Delta t'$, and $\Delta x'$ so we can calculate $\Delta\tau$ from $\Delta t'$ and $\Delta x'$ without knowing v.

This problem is analogous to a familiar one in Euclidean geometry. We consider a rotation by angle θ between two frames (Figure 4.5). In the frame O, the point P has coordinates (x, y) and in O ' it has coordinates (x', y'). They are related by a rotation,

$$x' = x\cos\theta + y\sin\theta$$

$$y' = y\cos\theta - x\sin\theta.$$

So, (x, y) is related to (x', y') by a linear transformation. We want to know what quantities are the same in both frames O and O'. In other words,

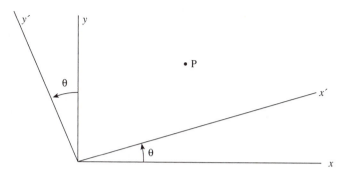

FIGURE 4.5 ▶

what quantities are preserved by rotation? Intrinsic geometrical relations are preserved, such as the angles between vectors and the length of vectors. For example, the distance P from the origin is independent of the choice of coordinates,

$$x^2 + y^2 = x'^2 + y'^2.$$

This relation is independent of θ—it is true for any two frames related by rotation.

What is the analogous result for Lorentz transformations? These transformations, as is clear from Minkowski's diagrams, do not preserve angles or Euclidean distances. So, $c^2(\Delta t')^2 + (\Delta x')^2$ will not work. Using $\Delta t' = \Delta\tau/\sqrt{1 - v^2/c^2}$ and $\Delta x' = -v\Delta t' = -v\Delta\tau/\sqrt{1 - v^2/c^2}$, we can calculate $c^2(\Delta t')^2 + (\Delta x')^2$ and check that it is not an invariant. But in so doing, we note that a change of sign does the trick:

$$c^2(\Delta t')^2 - (\Delta x')^2 = (\Delta\tau)^2 \frac{c^2}{1 - v^2/c^2} - (\Delta\tau)^2 \frac{v^2}{1 - v^2/c^2} = c^2(\Delta\tau)^2.$$

We learn from this that we can tell time in frame S by using measurements in another inertial frame S',

$$c\Delta\tau = \sqrt{(c\Delta t')^2 - (\Delta x')^2}. \tag{4.7}$$

We also learn that $(c\Delta t)^2 - (\Delta x)^2$ is an invariant (i.e., it is the same in all inertial frames), and so is called a Lorentz invariant.

This result is more general than our discussion so far. Take two events, one at $P_1 = (ct_1, x_1, y_1, z_1)$ and the second at $P_2 = (ct_2, x_2, y_2, z_2)$. Then, defining the differences $c\Delta t = ct_2 - ct_1$, and so on, we can check using the Lorentz transformation formulas Eqs. (4.2a) and (4.2b) that

$$(\Delta s)^2 \equiv (c\Delta t)^2 - (\Delta x)^2 - (\Delta y)^2 - (\Delta z)^2$$

is the same in all other frames. Unlike the invariant distance of Euclidean space, $x^2 + y^2 + z^2$, this invariant interval can be either positive, negative, or zero.

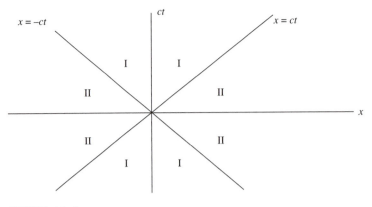

FIGURE 4.6 ▶

Let us illustrate these three possibilities and choose $P_1 = (0, 0, 0, 0)$ and $P_2 = (ct, x, 0, 0)$ for easy visualization. Clearly if P_1 and P_2 are connected by a light ray,

$$(\Delta s)^2 = 0 \qquad \text{(light-like)}.$$

This light ray separates our space-time plot into a region I where $(\Delta s)^2 > 0$ and a region II where $(\Delta s)^2 < 0$ (Figure 4.6). For P_2 in region I, we can boost to a frame where P_1 and P_2 occur at the same x'. In other words, the two events could be the ticks on a clock at rest in a frame S'. In that frame, $\Delta x' = 0$ and $\Delta t'$ is just the proper time $\Delta \tau$. Similarly, if P_2 lies in region II, we can boost to a frame S' where P_1 and P_2 occur simultaneously, $\Delta t' = 0$, so $-(\Delta s)^2$ then represents the proper length squared of a rod extending from P_1 to P_2. It is conventional to label region I time-like because $(\Delta s)^2 > 0$ and to label region II space-like because $(\Delta s)^2 < 0$.

Another important invariant distinction between regions I and II involves causality. If the event P_2 lies in region I for $t > 0$, one can find a frame S' where it occurs at the same x' as P_1 but at a later time t'. Clearly, then, an action at P_1 could cause a result at P_2. We say that the space-time points P_1 and P_2 are causally connected and that P_2 lies in P_1's "forward light cone." Similarly, if P_2 lies in the region I for $t < 0$, then an action at P_2 could cause a result at P_1 and the points are again causally connected. However, if P_2 lies in region II, one can find a physical frame S' where P_1 and P_2 are simultaneous. There is no way for them to communicate without inventing a signal that violates the speed limit c, and so they are not causally connected. In fact, we can also find a frame S'' where P_2 occurs *before* P_1, as shown in Figure 4.7. Space-time points separated by a negative interval $(\Delta s)^2$ do not have a unique time ordering and cannot influence one another. The region II is sometimes called "elsewhere" because of this.

We return to some of these notions when we discuss dynamics, a topic in which causality plays a particularly important role.

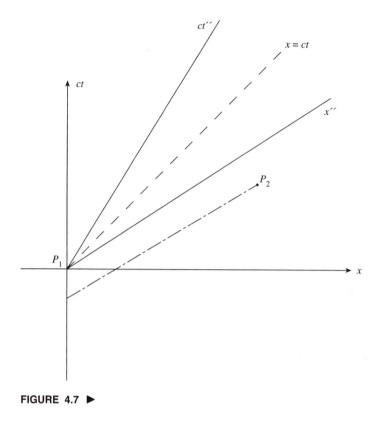

FIGURE 4.7 ▶

▶ PROBLEMS ▶

4-1 Two events occur at the same place in the frame S and are separated by a time interval of 5 s. What is the spatial separation between these two events in frame S′ in which the events are separated by a time interval of 7 s? Frame S′ moves at a constant velocity along the x direction of frame S.

4-2 Two events occur at the same time in the frame S and are separated by a distance of 2 km along the x axis. What is the time difference between these two events in frame S′ in which their spatial separation is 4 km? Frame S′ moves at a constant velocity along the x direction of frame S.

4-3 An event occurs at $x′ = 100$ m, $t′ = 9 \cdot 10^{-8}$ s in frame S′. If this frame moves with velocity $4c/5$ along the x axis of frame S, what are the space-time coordinates of the event in frame S?

4-4 The space and time coordinates of two events are measured in frame S to be

Event 1: $x_1 = L$, $t_1 = L/c$
Event 2: $x_2 = 2L$, $t_2 = L/2c$

    (a) Find the velocity of a frame S' in which both events occur at the same time.

    (b) What is the time t' that both events occur in S'?

4-5 Frame S' has a speed of $0.8c$ relative to S.

    (a) An event occurs at $t = 5 \cdot 10^{-7}$ s and $x = 100$ m in frame S. Where and when does it occur in S'?

    (b) If another event occurs at $t = 7 \cdot 10^{-7}$ s and $x = 50$ m, what is the time interval between the events in frame S'?

4-6 In parts (b) and (c) of this problem, we measure the times and locations of events by looking at them. In other words, instead of recording locations and times by being at the event, the event will send us a light signal that takes some additional time to propagate to us. This extra time must be taken into account accordingly. Astronomical data are of this type. For example, we see the Sun approximately 8 minutes in its past.

    Suppose a meter stick, aligned along the x direction, moves with velocity $0.8c$ and its midpoint passes through the origin at $t = 0$. Let there be an observer at $x = 0$, $y = 1$ m, $z = 0$.

    (a) Where in the observer's frame are the end points of the meter stick at $t = 0$?

    (b) When does the observer see the midpoint pass through the origin?

    (c) Where do the end points appear to be at this time, as seen by the observer?

4-7 Frame S' has velocity $0.6c$ relative to frame S. At $t' = 10^{-7}$ s a particle leaves the point $x' = 12$ m, traveling in the negative x' direction with a velocity $u' = -c/3$. It is brought to rest suddenly at time $t' = 3 \cdot 10^{-7}$ s.

    (a) What was the velocity of the particle as measured in the frame S?

    (b) How far did it travel as measured in the frame S?

4-8 Frame S' has velocity v relative to frame S. At time $t = 0$ a light ray leaves the origin of S, traveling at a 45° angle with the x axis.

    (a) What angle does the light ray make with respect to the x' axis in the frame S'?

    (b) Repeat part (a) replacing the light ray with a particle of mass m and speed u.

    (c) Repeat part (a) replacing the light ray with a rod that is stationary in frame S.

4-9 The Fizeau Experiment. Light propagates more slowly through a material medium than through a vacuum. If v_m is the speed of light in a medium and the medium is moving with respect to the frame S at the velocity v, find an expression for the velocity of light in the frame S, assuming that the light ray propagates in the same direction as v. Show that when v is much smaller than c, this general expression reduces to $v_m + v(1 - v_m^2/c^2)$, to leading order in v. (This expression was tested by H. Fizeau in the mid-nineteenth century using interferometry techniques.)

4-10 The Čerenkov Radiation. Although light travels at the speed limit in empty
space, its speed is diminished as it propagates through materials, as discussed in the Fizeau experiment. The process by which light is slowed is
complicated—the waves scatter from the atoms making up the material
and a light wave can lose speed and coherence in the process. We learn
about these scattering processes in quantum mechanics. Nonetheless, light
may not even be the fastest racer in a particular material. When a charged
particle moves through the material faster than the speed of light in that
material, it radiates coherent light in a cone that trails behind it, as shown
in the figure. Note in the figure that the charged particle is shown to radiate
light at several points, and the resulting wave front of Čerenkov radiation,
the conical surface BC, is perpendicular to the direction of propagation of
the light rays.

(a) Show that the half-angle of the cone of Čerenkov radiation, ϕ, is
$\cos \phi = v_m/v$, where v_m is the speed of light in the material and v is the
speed of the charged particle. (In high energy physics and cosmic ray
experiments, Čerenkov radiation and this formula are used to measure
the speeds of exotic charged particles.)

(b) Compare this phenomenon to these well-known ones: When a speed
boat travels at high velocity over the water, it leaves a sharp wake
behind it; and when a jet plane travels faster than the speed of sound,
it makes a sonic boom.

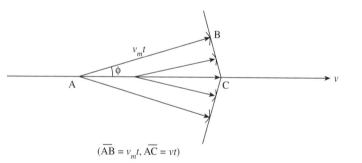

$(\overline{AB} = v_m t, \overline{AC} = vt)$

4-11 Stellar Aberration. Imagine that Earth were at rest with respect to a distant
star and that an Earthbound telescope had to be pointed at an angle θ above
the horizon to view the star. Now suppose that there is a relative velocity
v between Earth and the star. The angle θ would change to θ'.

(a) Show that θ' is given by $\cos \theta' = (\cos \theta + v/c)/[1 + (v/c) \cos \theta]$.

(b) If v/c is very small, show that the formula in part (a) reduces to
$\cos \theta' \approx \cos \theta + (v/c) \sin^2 \theta$.

(c) Because the difference between θ and θ' is very small under the conditions of part (b), it is convenient to introduce the angle $\alpha \equiv \theta' - \theta$.
Show that part (b) reduces to the prediction $\alpha \approx -(v/c) \sin \theta$.

This result is called stellar aberration and is significant in astronomical observations. It is particularly important when we compare the position of a star in the sky in winter to its position in summer, because the velocity of Earth reverses relative to the star every 6 months.

4-12 Headlight Effect. Suppose that a car's headlight sends out light rays into a forward hemisphere. Use the result of Problem 4.11a to deduce that the maximum angle of the light with respect to the line of motion of the car is $\cos \theta' = v/c$ in the rest frame of the road. Explain. (This result is called the headlight effect.)

4-13 A frame S' moves along the x axis of frame S at a velocity v. A particle moves with a velocity $v'_x = 0$ and $v'_y \neq 0$ in the S' frame. (a) What are v_x and v_y in the frame S? (b) Why are v'_y and v_y different?

4-14 A frame S' moves along the x axis of frame S at a velocity v. In frame S there is a meter stick parallel to the x axis and moving in the y direction with a velocity v_y, as shown in the figure. The center of the meter stick passes the point $x = y = x' = y' = 0$ at $t = t' = 0$.

(a) Argue, on the basis of the relativity of simultaneity, and without calculation, that the meter stick is measured as tilted upward in the positive x' direction in the frame S'!

(b) Calculate the angle of the tilt, ϕ', in frame S', as shown in the figure, by answering several easy questions. Where and when does the right end of the meter stick cross the x axis as observed in the frame S'? (You might answer this question first in the frame S and then transform this information to the frame S'.). Referring to Problem 4.13 for the velocity of the meter stick in the frame S', determine where the right end of the meter stick is at time $t' = 0$ when the center is at the origin. Your formula for ϕ' follows from this. [Ref: R. Shaw, *Am. J. Physics* 30, 72 (1972).]

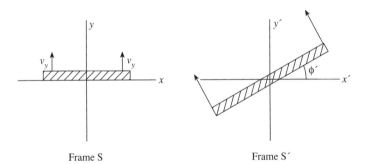

        Frame S   Frame S'

► Chapter 5

Illustrations and Problems in Space-Time Measurements

Let us apply the principles we have learned so far to some interesting problems. Some of these problems are solved in several different ways to illustrate their pitfalls and strategies.

5.1 A Spaceship Rendevous

A spaceship of length 100 m has a radio receiver in its nose [1]. It travels at a velocity $v = 0.6c$ relative to a space station. A radio pulse is emitted from the space station just as the tail of the spaceship passes by the transmitter.

a. How far from the space station is the nose of the spaceship when it receives the radio signal?

b. How long did it take the radio pulse to reach the nose of the spaceship from the perspective of the space station?

c. How long did it take the radio pulse to reach the nose of the spaceship from the perspective of the spaceship?

Consider Figure 5.1 showing the emission and reception of the radio pulse. From the perspective of the space station, the radio pulse travels a distance ct and goes a distance vt plus the length of the spaceship in its (the space station's) rest frame. This distance is 100 m/γ. So,

$$ct = \frac{100}{\gamma} + vt.$$

FIGURE 5.1 ▶

The relevant distance is 100 m/γ because the spaceship's proper length, 100 m, is being measured in a frame at relative velocity v. Since $v = 0.6c$, we compute $\gamma = 1/\sqrt{1 - v^2/c^2} = 5/4$ and solve for ct, finding $ct = 200$ m. This answers part a. For part b, $t = 200$ m/$c = 200/3 \cdot 10^8 \approx 6.67 \cdot 10^{-7}$ s. Now we need the time elapsed in the spaceship's frame t'. Over this time interval, light travels from the rear of the spaceship to its nose, a distance of 100 m, at velocity c. So, $t' = 100/c = 3.37 \cdot 10^{-7}$ s, which answers part c.

Where did we use the principles of relativity? First, we had Lorentz contraction in part a, and second, in part c we used the fact that light travels at the speed limit c in any inertial frame. Easy problem—but wait, why aren't the times t and t' related by time dilation, a factor of $\gamma = 5/4$? Instead, we found in a most elementary way that $t' = t/2$. This is puzzling, until you notice that the events, the emission and absorption of the radio pulse, occur at *separate* points in the spaceship frame. From the perspective of the space station, a clock in the tail of the spaceship and one in the nose are not synchronized, and this time difference contributes to t'.

To see the effect, let us redo the problem by plugging into our Lorentz transformation formulas. First, we need the space-time coordinates of the two events in the spaceship frame. A radio pulse is emitted at $t'_1 = x'_1 = 0$ and is received in the nose of the spaceship at $x'_2 = 100$ m, $t'_2 = x'_2/c = 100$ m/$(3 \cdot 10^8)$ m/s $= 3.33 \cdot 10^{-7}$ s. So, in the space station's frame the second event occurs at

$$x_2 = \gamma(x'_2 + vt'_2) = \frac{5}{4}(100 + 0.6 \cdot 100) = \frac{5}{4} \cdot 1.6 \cdot 100 = 200 \, \text{m}$$

as we calculated before from the space station's perspective. The reception of the radio pulse occurs at

$$t_2 = \gamma\left(t'_2 + \frac{v}{c^2}x'_2\right) = \frac{5}{4}\left(\frac{100}{c} + 0.6\frac{100}{c}\right) = \frac{5}{4} \cdot 1.6 \cdot \frac{100}{c} = \frac{200}{c} = 6.67 \cdot 10^{-7} \, \text{s}$$

as before. Here we see the relativity of simultaneity at work. The formula states that t_2 is *not* time dilation applied to t'_2; it is time dilation applied to $t'_2 + (v/c^2)x'_2$. We recognize this as the time that the space station says passes on a clock at the nose of the spaceship between the two events. Multiplying by γ gives the time in the space station's frame. The term vx'_2/c^2 is the extra time that the space station says must pass on the clock at the nose

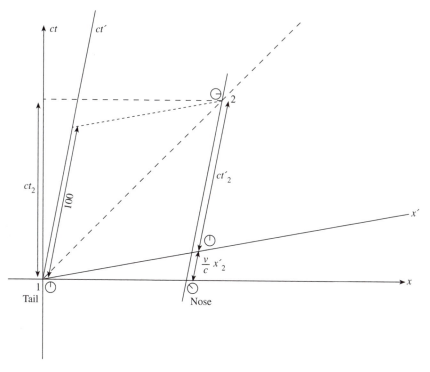

FIGURE 5.2 ▶

of the spaceship because that clock is vx_2'/c^2 seconds *behind* the clock at the tail.

All this becomes evident if we make a Minkowski diagram of the situation (see Figure 5.2). Note that the clocks in the tail and the nose are synchronized in the spaceship frame, S'. We see that, from the point of view of the space station, the clock in the nose is vx_2'/c^2 behind the clock in the tail and that ct_2 is related by time dilation to the *sum* of $ct_2' = 100$ and $vx_2'/c = 0.6 \cdot 100$ m.

5.2 A Hole in the Ice

A relativistic skater with 15-inch-long blades on his skates travels at $v = 0.8c$ over an icy surface. There is a hole in the ice of diameter 10 inches before the finish line. The skater decides to skate over the hole to win the race. He figures that this is safe because the diameter of the hole is Lorentz contracted in his frame to $10/\gamma = 10/(5/3) = 6$ inches, which is less than the length of his blade. A judge at the finish line sees the skater approaching and tries to wave him off because in *her* frame the length of the skater's blade

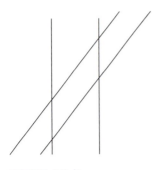

FIGURE 5.3 ▶

is $15/\gamma = 15 \cdot (3/5) = 9$ inches, so the entire blade will fit into the hole and the skater will fall through the ice and be injured. Which person is right? We better get the right answer because a serious accident lies in the balance!

This is a paradox of relativity almost as famous as the Twin Paradox. Let's describe it in a Minkowski diagram in which frame S is the rest frame of the ice (or the judge) and S′ is the rest frame of the skater's blade (or the speed skater). The world lines of the edges of the hole and the ends of the blade are shown in Figure 5.3 [2].

Now we want to describe these events in various frames where we lay down lines of constant time at various angles. From the perspective of the ice (or the judge), the lines of constant time t are horizontals as shown in Figure 5.4. At t_1, the front of the blade reaches the edge of the hole. At t_2, the back of the blade does the same. Between t_2 and t_3, the entire blade is inside the hole. At time t_3, the front end of the blade hits the other edge of the hole and the skater falls! Figure 5.4 ignores this and shows the blade emerging from the hole, and finally, at t_4, the back end of the blade comes out of the hole.

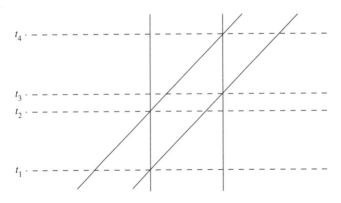

FIGURE 5.4 ▶

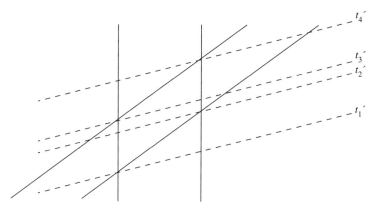

t_4'

t_3'

t_2'

t_1'

FIGURE 5.5 ▶

From the perspective of the skater, things are quite different because his lines of constant time t' are inclined to the horizontals by an angle $\tan \theta = v/c = 0.8$. These lines are shown in Figure 5.5. Inspecting this diagram, we see that the skater's impressions were also correct—at no time t' are both ends of his blade inside the hole—and even the time order of the events is different than that given in frame S. In S', the front of the blade emerges from the hole (at t_2') *before* the back end comes in (at t_3'). We can easily check that the invariant interval $(\Delta s)^2$ for the two events, the back end of the skater's blade reaches the left edge of the hole and the front end of the blade reaches the right edge of the hole, is negative. So, the events are not casually connected and their time ordering is frame dependent. This is obvious from the Minkowski diagram. We could also imagine that the back end of the skate emits a light pulse when it reaches the left edge of the hole. The pulse reaches the right end of the hole after a time interval of $10/c$ in the rest frame of the hole. The front end of the skate reaches the right edge of the hole after a much shorter time interval, $(10 - 9)/v = 5/4c$.

So, does the skater fall or not? Yes, he falls, and our discussion of causality should remind us of a flaw in the skater's description of his blade. When he claimed that his skate could not fit into the hole in the ice, he was thinking nonrelativistically, where rigid bodies exist. But in a relativistic world where information cannot travel faster than light, the blade is not rigid and the front end is pulled down by the force of gravity when $t_2' > t' > t_1'$. It is easy to check that if the front end of the skater's blade sent a radio pulse to the back end at time t_1', when the front end reaches the left edge of the hole, the pulse does not reach the back end of the blade until well after the front end has hit the right edge of the hole. In other words, the back end of the blade doesn't even know that there is a hole in the ice until after the accident!

This example suggests that Lorentz invariant methods would be the best tools to analyze dynamic, relativistic problems involving extended objects. Our frame-dependent discussion comes complete with biases and arbitrariness that can be misleading. In fact, high energy theorists who study relativistic strings as a framework for grand unification almost exclusively use invariant methods so that they concentrate on the physical content of processes involving strings—those aspects that all observers in any inertial frame can agree on.

5.3   A Velocity Greater Than the Speed Limit?

Consider three frames of reference. An observer O sees an observer S moving to her right at speed $v = 0.8c$ and another observer S' moving to her left at the same rate.

a. The observer O says, "The velocity difference between S and S' is $1.6c$, so the distance between them, $d = 1.6ct$, is growing at a rate in excess of the speed of light c." Is this statement correct? Does it contradict Postulate 2 of special relativity?

b. If S' measures the velocity of S, what does he find?

c. If a transmitter at rest with S' broadcasts light with a frequency $v' = 10$ cps, what frequency is measured by O and what frequency is measured by S?

Begin with a picture of the three observers shown in Figure 5.6. There is no doubt that O measures the distance between S and S' to be $d = 1.6ct$. This is just arithmetic. But $1.6c$ does not represent the speed of one object relative to another, so there is no contradiction with relativity. No physical object is moving in an inertial frame with a speed in excess of c. We see this explicitly by considering the next part of the problem. Using our relativistic addition of velocities formula, the speed of S in the frame of S' consists of two pieces: the speed of S in the rest frame of O, v, and the speed of O relative to S', v again. So, the speed of S measured by S' is

$$u = \frac{(v+v)}{(1+\frac{v^2}{c^2})} = \frac{1.6c}{(1+0.8)^2} = \frac{1.6c}{1.64} = 0.9756c,$$

which is, indeed, slightly less than c. All is well.

FIGURE 5.6 ▶

It is amusing to note in passing that the situation depicted here approximates several natural phenomena studied in astrophysical research journals. Some researchers have presented analyses of superluminal jets of gases streaming from spinning black holes, much as S and S′ are racing away from O. When observed through an Earth-bound telescope, the jets appear to be diverging at a rate exceeding c. These and other optical effects are sometimes raised in the popular science literature as threats to relativity. They don't last long.

Back to our problem. The relation between the frequency ν' and the frequency ν_o, observed by O, is given by the Doppler formula,

$$\nu_o = \sqrt{\frac{1-v/c}{1+v/c}}\,\nu' = \sqrt{\frac{1-0.8}{1+0.8}}\,\nu' = \frac{10}{3}\ \text{cps.}$$

Finally, we can calculate the frequency ν measured by S by noting that he observes the wave train in O's reference frame as Doppler shifted by another factor of $\sqrt{(1-v/c)/(1+v/c)}$ because the relative speed between O and S is v,

$$\nu = \sqrt{\frac{1-v/c}{1+v/c}}\,\nu_o = \left(\frac{1-v/c}{1+v/c}\right)\nu' = \frac{1-0.8}{1+0.8}\cdot 10 = \frac{10}{9}\ \text{cps.}$$

Alternatively we could calculate ν directly in terms of ν' by again applying the Doppler formula, but using the relative velocity u between S and S′. Call $\beta = v/c$, so

$$\nu = \sqrt{\frac{1-u/c}{1+u/c}}\,\nu' = \sqrt{\frac{1-2\beta/(1+\beta^2)}{1+2\beta/(1+\beta^2)}}\,\nu' = \left(\frac{1-v/c}{1+v/c}\right)\nu',$$

where we did some algebra in the previous step, and found agreement with the result.

▶ Problems ▶

5-1. **The Scissors Paradox.** A long straight rod that is inclined at an angle θ to the x axis has a velocity v downward as shown in the figure.

   (a) Derive a formula for the speed v_r of the intersection of the rod with the x axis in terms of v and θ.
   (b) Can you choose values for v and θ so that v_r is greater than the speed limit c?
   (c) Can you transmit information along the x axis at speeds greater than c using this trick?

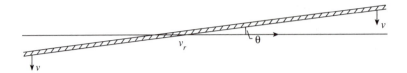

5-2. Tricks Your Eyes Can Play. Consider a rocket that goes from point 1 to point 2, a total distance d, in time t at velocity v. At both points it radiates a pulse of light, and these pulses are detected by an observer O at rest to the right of point 2, as shown in the figure.

   (a) Show that the time interval between the reception of the pulses at O is $T = [1 - (v/c)]t$, where t is the time it took the rocket to travel from point 1 to point 2, $d = vt$.

   (b) Because the light pulses provide observer O with two images of the rocket, one at point 1 and a second at point 2, a distance d to the right of point 1, argue that she sees the rocket moving with an apparent speed $d/T = v/[1 - (v/c)]$, which is greater than c, whenever $c/2 < v < c$!

   (c) Suppose that there is a stationary observer O' to the left of point 1. Find a formula for the apparent speed that he attributes to the rocket. What is its maximum value?

   This problem illustrates the visual effects that occur in astrophysical observations of moving and exploding stars. Interpreting astronomical data can get pretty tricky! If you pursue astrophysics, you will see more such effects in more realistic settings. This problem also illustrates why we are so particular about how measurements are made in relativity— why we phrase measurements in terms of events, specifying the time an event occurs on a clock at that exact spot.

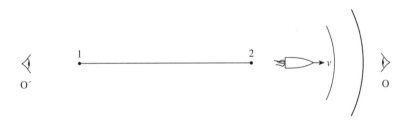

5-3. A Project to Elucidate the Physics of Section 5.2. Two long parallel conducting rails are open at one end, but connected electrically at the other end through a lamp and battery, as shown in the figure. An H-shaped slider, whose vertical pieces are made of copper but whose horizontal piece is an insulator, moves along the rails. Let the length of the slider be 2 m, matching the rest length of the region AB, and let the height of the slider match the distance between the rails, except in the special region AB where the rails are further apart. Suppose that the H-shaped slider moves at a large velocity, so that it contracts by a factor of 2 relative to the region AB. In the rest frame of the rails, an observer concludes that there is a period of time when the slider does not complete the circuit between the battery and the lamp, so the lamp must go dark for a moment. However, from the perspective of the slider, the region AB is contracted by a factor of 2, so an observer on the slider predicts that the circuit is always completed and the lamp will always shine. Analyze this system as we analyzed the ice skate and the hole problem in Section 5.2 and resolve the paradox. Does the lamp go dark for an instant or not?

This project assumes that the reader knows how signals, current and voltages, propagate along a wire. The time delay between the arrival of the right end of the slider at point *B* and the transmission of that information from point *B* back to point *A* on the wire is an important part of the resolution of this puzzle. (It is the analog of the time it takes the back end of the skate to know that the front end crashed into the far side of the hole.) This challenging problem can be used as a fascinating research project involving the reading and mastery of the material in the reference. [Ref: G. P. Sastry, *Am. J. Physics* 55, 943 (1987).]

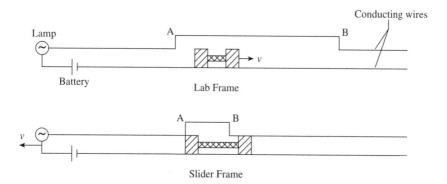

Lab Frame

Slider Frame

Chapter 6

Relativistic Dynamics

6.1 | Energy, Light, and $E = mc^2$

We have spent all our time so far discussing how measurements of space-time coordinates transform between inertial frames in such a way that there is a universal speed limit. Has this been an empty exercise? We need to see that we can write down a scheme of relativistic dynamics that satisfies Postulate 1 of relativity—that the relative speed between inertial frames has no material influence on the dynamics in either frame. We know that the transformation laws between inertial frames in a Newtonian world, the Galilean transformations we have discussed in previous sections, are consistent with Newton's laws of motion. Now we need the relativistic laws of motion to underpin the Lorentz transformations.

The hallmarks of Newtonian mechanics are the Second Law (that force equals inertia times accleration) and the Third Law (that action equals reaction; i.e., if I push on you, you push on me in an equal and opposite fashion). These results lead to momentum conservation. The Galilean transformations then imply that if momentum conservation is true in one inertial frame, it is true in them all. These transformations implement the idea that all inertial frames are equivalent.

Let us briefly review these Newtonian developments before moving on to relativistic dynamics. The logic here is important. Consider two point particles of mass m_1 and m_2 and let them collide in an inertial frame where, by definition, there is no external force. So, if $\mathbf{f}_{12}$ is the force that particle 1 exerts on particle 2 and $\mathbf{f}_{21}$ is the force that particle 2 exerts on particle 1, Newton's Second and Third Laws imply

$$\frac{d}{dt}(\mathbf{p}_1 + \mathbf{p}_2) = \mathbf{f}_{21} + \mathbf{f}_{12} = 0,$$

where $\mathbf{p}_1$ is the momentum of particle 1, $\mathbf{p}_1 = m_1\mathbf{v}_1$, and $\mathbf{p}_2 = m_2\mathbf{v}_2$. So, the total momentum is conserved, $m_1\mathbf{v}_1 + m_2\mathbf{v}_2 = \text{constant}$. This means that the center of mass $\mathbf{R} = (m_1\mathbf{x}_1 + m_2\mathbf{x}_2)/(m_1 + m_2)$ travels with a constant velocity, $(d/dt)\mathbf{R} = \text{constant}$, and we can boost to a frame where $\mathbf{R}$ is the origin and remains there forever. But we can also view the collision in another frame S′, which moves with velocity $\mathbf{v}$ with respect to frame S. By Galilean invariance, we have the nonrelativistic rule of addition of velocities, so the velocities of the particles in frame S′ read

$$\mathbf{v}'_1 = \mathbf{v}_1 - \mathbf{v}, \qquad \mathbf{v}'_2 = \mathbf{v}_2 - \mathbf{v}.$$

So, in the frame S′,

$$\mathbf{p}'_1 = m_1\mathbf{v}'_1 = m_1\mathbf{v}_1 - m_1\mathbf{v} = \mathbf{p}_1 - m_1\mathbf{v}$$
$$\mathbf{p}'_2 = m_2\mathbf{v}'_2 = m_1\mathbf{v}_2 - m_2\mathbf{v} = \mathbf{p}_2 - m_2\mathbf{v}.$$

So,

$$\frac{d}{dt'}(\mathbf{p}'_1 + \mathbf{p}'_2) = \frac{d}{dt}(\mathbf{p}_1 + \mathbf{p}_2) = 0$$

because $\mathbf{v}$ is a constant. So, the same dynamics holds in either frame and, if we have momentum conservation in one inertial frame, we have it in all of them.

In short, Newtonian dynamics works together with Galilean invariance to assure that physics satisfies Newton's version of relativity. The extension of this consistency to the relativistic domain will take some work because, although relativistic dynamics must reduce to Newtonian dynamics when all the relative velocities involved are small compared to the speed limit c, Lorentz transformations are more intricate than Galilean transformations. We can anticipate that the Newtonian notion of momentum $\mathbf{p} = m\mathbf{v}$ will not work in Einstein's world because the frame independence of momentum conservation used the addition of velocities rule, which is not true relativistically.

So, we must follow Einstein and rethink the notions of momentum, mass, and energy. To begin, let us present one of Einstein's famous thought experiments that shows that relativistic energy and inertia must be unified into one concept in the new dynamics.

Consider a box of mass M and length L and suppose that radiant energy (light) E is emitted from one end and absorbed by the other (Figure 6.1). We need to know one fact about light, which we will understand in more detail later: if a wave carries energy E, it also carries momentum p and they are related by $E = pc$. So, when light is emitted from the left end of the box, as shown in the top half of Figure 6.1, the box recoils with a momentum $-E/c$. The box then moves to the left with velocity $v = -E/Mc$, supposing that M is so large that v is very small, $v \ll c$, and the box's

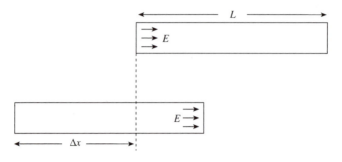

FIGURE 6.1 ▶

motion is well described by Newtonian nonrelativistic considerations. Then, the light reaches the right end of the box in time $\Delta t = L/c$ and is absorbed. Now, the box should be at rest again, but apparently it has moved by a distance $\Delta x = v\Delta t = -EL/Mc^2$! But this is crazy—the center of mass of the system can't move because there are no external forces to push it. If our expression for Δx were correct, we could repeat this process as often as we wished and move the box as far as we liked to the left. Something is wrong! M is large and Newtonian mechanics should work to describe it.

It must be that when energy E moves from one end of the box to the other it must deposit some mass m on the right-hand end of the box, so the center of mass of the heavy box doesn't move. How much should m be to do the trick? Certainly $m \ll M$. The condition that $\Delta \bar{x}$ (the change in the position of the center of mass) vanishes reads

$$\Delta \bar{x} = 0 = mL + M\Delta x.$$

Solving for m,

$$m = -\frac{M}{L}\Delta x = \frac{M}{L}\frac{EL}{Mc^2} = E/c^2.$$

In other words,

$$E = mc^2. \tag{6.1}$$

This means that the energy carried by the light wave results in a mass increase of $m = E/c^2$ when it is absorbed as heat on the end of the box. Because c is so large, m is extraordinarily tiny, and this thought experiment is not practical. However, in the realm of nuclear and high energy physics, dramatic illustrations of Eq. (6.1) can be found. Some are discussed later.

Eq. (6.1) is one of the most famous equations in physics. When we study elementary particle collisions later, we shall see that it predicts that we can convert the rest mass of heavy particles into the kinetic energy of lighter ones. The equation underlies nuclear power.

Because mass can be converted into energy and vice versa, according to Eq. (6.1), one of the sacred conservation laws of Newton must be modified. That conservation law states that mass, as well as total momentum, is conserved in a collision of particles in an inertial frame,

$$\mathbf{p}_1 + \mathbf{p}_2 = \mathbf{p}_1' + \mathbf{p}_2'$$
$$m_1 = m_1', \qquad m_2 = m_2'.$$

According to Newton, there is separate overall momentum conservation and mass conservation for each elementary particle. Our next task is to find the relativistic generalization of these statements.

6.2 Patching up Newtonian Dynamics—Relativistic Momentum and Energy

Can we invent a formula for momentum, a relativistic analog to $\mathbf{p} = m\mathbf{v}$, so that relativistic momentum is conserved in the collision of particles in an inertial frame and the conservation law is truly relativistic (i.e., it holds in all inertial frames if it holds in one)? Let's take a simple collision and come to an answer with a mininum of algebra.

First, consider a Newtonian inelastic collision, shown in Figure 6.2. Initially there is a particle of mass m at rest in the lab, and a particle of the same mass but with velocity u collides and sticks to it. The final composite particle of mass $2m$ recoils with velocity $\bar{u}$, determined by momentum conservation,

$$mu = 2m\bar{u}.$$

So,

$$\bar{u} = \frac{1}{2}u$$

Because the initial kinetic energy is $T_i^{\mathrm{NR}} = (1/2)mu^2$ (NR denotes nonrelativistic, or Newtonian) and the final kinetic energy is $T_f^{\mathrm{NR}} = (1/2)(2m)\bar{u}^2 = (1/4)mu^2$, heat Q is generated,

$$Q = T_i^{\mathrm{NR}} - T_f^{\mathrm{NR}} = \frac{1}{4}mu^2.$$

In this discussion we assumed that the mass of the composite particle was twice the mass of its parts. Actually, the additivity of mass is forced

$\bigcirc\!\!\longrightarrow u \qquad \bigcirc \qquad\qquad\qquad \bigcirc\!\!\bigcirc\!\!\longrightarrow \bar{u}$

Before   After

FIGURE 6.2 ▶

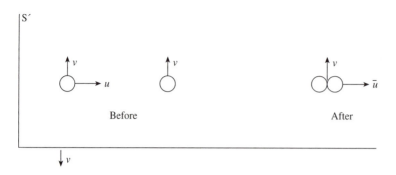

FIGURE 6.3 ▶

on us by Postulate 1 and the form of the Newtonian momentum. To see this, call the mass of the composite particle M_o and view the collision in a frame S', which has a *transverse* velocity $\mathbf{v}$ (Figure 6.3).

Conservation of transverse momentum reads

$$m\mathbf{v} + m\mathbf{v} = M_c\mathbf{v}.$$

So, we read off $M_c = 2m$, as expected. What is the point of this discussion? Conservation of momentum must hold in any inertial frame and it will do so only if masses add in a Newtonian world.

Now we want a *relativistic* description of this process that will satisfy Postulate 1—the law of conservation of momentum must be the same in all frames [7]. Our task is to find an expression for the relativistic momentum that accomplishes this. The expression must reduce to its nonrelativistic cousin, $\mathbf{p} = m\mathbf{v}$, when the velocity is small compared to the speed limit c. The beam particle in the initial state in Figure 6.2 has relativistic momentum, which must have the mathematical form

$$\mathbf{p} = f(u/c)m\mathbf{u}.$$

This expression should accommodate what we know about the momentum. A study of the collision should determine the function f. Why have we written this form for $\mathbf{p}$? First, we have chosen $\mathbf{p}$ to point in the direction $\mathbf{u}$ because that is the only vector in the problem. We have taken $\mathbf{p}$ to be proportional to the inertia m so its dimensions are correct. Finally, there is the possibility that f depends on the dimensionless ratio u/c, the ratio of the magnitude of $\mathbf{u}$ to the speed limit. The dimensionless function f incorporates features of $\mathbf{p}$ that we need to determine. It is conventional to write $m(u) \equiv f(u/c)m$ and call $m(u)$ the relativistic mass.

Because there are no external forces, we require that relativistic momentum be conserved,

$$m(u)\mathbf{u} = M(\bar{u})\bar{\mathbf{u}}, \tag{6.2}$$

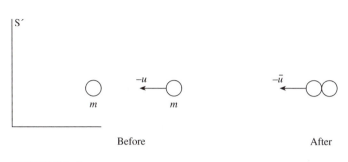

Before   After

FIGURE 6.4 ▶

where $M(\bar{u})$ will be determined by momentum conservation and Postulate 1. We do *not* assume $M(\bar{u}) = 2m(\bar{u})$, as nonrelativistic reasoning might suggest. In fact, this is *not* true!

To see how Postulate 1 constrains all the apparent arbitrariness here, view the collision in a frame S′, where the beam particle of Figure 6.2 is at rest as in Figure 6.4. In this frame the collision is just turned around. But frame S′ is obtained from frame S by a boost (Figure 6.5). So, the velocity of the composite particle, as measured in frame S, must also be given by an application of the addition of velocities formula,

$$\bar{u} = \frac{-\bar{u} + u}{1 - u\bar{u}/c^2},$$

which we can solve for u,

$$u = \frac{2\bar{u}}{1 + \bar{u}^2/c^2}. \tag{6.3}$$

Is this result reasonable? Well, if $\bar{u} \ll c$, then $u \approx 2\bar{u}$, and we are back to the Newtonian result we had before. All is well, so far.

To determine $M(\bar{u})$ by Postulate 1 of relativity, view the collision in a frame with a small transverse velocity in the $-y$ direction (Figure 6.6). In the frame S′, the beam particle (see "Before" in Figure 6.7) then has $v_y' = v$,

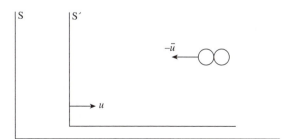

FIGURE 6.5 ▶

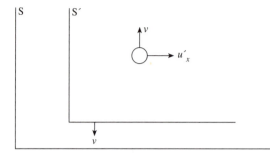

FIGURE 6.6 ▶

the transverse velocity between the frames. The beam particle had velocity u in the x direction in the lab frame, so in frame S'

$$u'_x = \frac{x'}{t'} = \frac{x}{\gamma(v)t} = \frac{1}{\gamma(v)}u,$$

where $\gamma(v) = 1/\sqrt{1 - v^2/c^2}$, using our time dilation formula. Now we can write momentum conservation for the collision in frame S' (Figure 6.7).

But in the frame S' the mass of the beam particle is a function of the length of the velocity $\mathbf{u}'$, which is $\mathbf{u}'^2 = v^2 + u'^2_x = v^2 + u^2(1 - v^2/c^2)$. The same remark applies to $\bar{\mathbf{u}}'$, the velocity of the composite particle, $\bar{\mathbf{u}}'^2 = v^2 + \bar{u}'^2_x = v^2 + \bar{u}^2(1 - v^2/c^2)$. So, conservation of the y component of momentum in frame S' reads

$$m(u')\mathbf{v} + m(v)\mathbf{v} = M(\bar{u}')\mathbf{v},$$

which gives the composite mass $M(\bar{u}')$. We only need this general formula for the special situation $v \to 0$,

$$m(u) + m(0) = M(\bar{u}). \tag{6.4}$$

This formula tells us how to add the relativistic masses of the colliding particles to make up the mass of the composite. It is a statement of mass conservation for relativistic systems where mass must depend on velocity. Without Eq. (6.4) we could not have conservation of relativistic momentum. Substituting into Eq. (6.2), momentum conservation in the lab frame S,

$$m(u)\mathbf{u} = M(\bar{u})\bar{\mathbf{u}} = (m(u) + m(0))\bar{\mathbf{u}},$$

Before   After

FIGURE 6.7 ▶

we can now determine the velocity dependence of the relativistic mass,

$$m(u) = m(0) \cdot \frac{\bar{u}}{u - \bar{u}}.$$

This result is useful when we express $\bar{u}$ in terms of u, the velocity of a beam particle in the lab frame. This can be done using Eq. (6.3), giving

$$\frac{\bar{u}}{u - \bar{u}} = \frac{\bar{u}}{\frac{2\bar{u}}{1 + \bar{u}^2/c^2} - \bar{u}} = \frac{1 + \bar{u}^2/c^2}{1 - \bar{u}^2/c^2}, \tag{6.5}$$

where we also did some algebra to get the last expression. The right-hand side of Eq. (6.5) must be written in terms of u. Using the identity

$$(1 - \bar{u}^2/c^2)^2 = (1 + \bar{u}^2/c^2)^2 - 4\bar{u}^2/c^2$$

we have

$$\left(\frac{1 - \bar{u}^2/c^2}{1 + \bar{u}^2/c^2}\right)^2 = 1 - \frac{4\bar{u}^2/c^2}{(1 + \bar{u}^2/c^2)^2} = 1 - u^2/c^2,$$

where we identified Eq. (6.3), $u = 2\bar{u}/(1 + \bar{u}^2/c^2)$, in the last step. So, we end with the elegant result,

$$\frac{\bar{u}}{u - \bar{u}} = \gamma(u),$$

which gives the relativistic mass

$$m(u) = \gamma(u)m(0) = m(0)/\sqrt{1 - u^2/c^2}.$$

In summary, we have the very important result for the relativistic momentum of a particle of rest mass $m(0) = m$ and velocity **u**,

$$\mathbf{p} = \gamma m \mathbf{u}, \qquad \gamma = 1/\sqrt{1 - u^2/c^2}. \tag{6.6}$$

What do we learn from all this? What are the special features of **p**? **p** has a factor of $\gamma = 1/\sqrt{1 - u^2/c^2}$, which crept in from the exercise in addition of relativistic velocities. The factor of γ, therefore, is forced on us by the way positions and times transform from one inertial frame to another.

Because Eq. (6.6) is much simpler than its derivation, we must find a more fundamental derivation for it. We will return to this later.

One of the satisfying features of Eq. (6.6) is that it is consistent with the speed limit c. Because γ grows without bound as u approaches c, the relativistic mass $m(u)$ and the relativistic momentum $\mathbf{p} = m(u)\mathbf{u}$ do so as well.

There is still more to be learned from our inelastic collision. Let's look at the mass of the composite particle in more detail. To do this, boost to yet another frame, the center of momentum frame where the total momentum vanishes. This frame is clearly obtained from the lab frame by a boost to

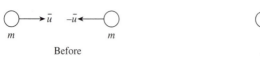

FIGURE 6.8 ▶

the left through velocity $\bar{u}$, just sufficient to bring the composite particle to rest. The collision is as shown in Figure 6.8.

Just as we found $m(u) + m(0) = M(\bar{u})$ by considering momentum conservation in the lab frame, we now find $m(\bar{u}) + m(-\bar{u}) = M(0)$. Because m depends only on the magnitude of its argument, $m(-\bar{u}) = m(\bar{u})$, we have, more simply,

$$M = 2m(\bar{u}) = \frac{2m}{\sqrt{1 - \bar{u}^2/c^2}}. \qquad (6.7)$$

This result replaces the Newtonian conservation law in which M would be just twice m. If $\bar{u}^2/c^2 \ll 1$, Eq. (6.7) will not deviate much from its Newtonian limit. Using the expansion from Appendix A,

$$\frac{1}{\sqrt{1 - \bar{u}^2/c^2}} \approx 1 + \frac{1}{2}\frac{\bar{u}^2}{c^2} + O\left(\frac{\bar{u}^4}{c^4}\right),$$

we have

$$M \approx 2m\left(1 + \frac{1}{2}\frac{\bar{u}^2}{c^2}\right) = 2m + 2 \cdot \frac{m\bar{u}^2}{2c^2}. \qquad (6.8)$$

Here we can identify the nonrelativistic kinetic energy $T^{NR} = m\bar{u}^2/2$. The total kinetic energy, $2 \cdot (m\bar{u}^2/2)$, is converted to heat Q in this inelastic collision, so Eq. (6.8) can be written

$$M \approx 2m + Q/c^2.$$

Identifying the inertial mass in this inelastic collision,

$$\Delta M \equiv M - 2m = Q/c^2,$$

we obtain Einstein's famous formula,

$$Q = \Delta Mc^2,$$

which states that heat Q is equivalent to rest mass ΔM through a conversion factor of c^2. This is the same result as in Section 6.1, obtained from a different perspective.

In order to assimilate all these statements about energy and mass, we define the relativistic energy to be

$$E = \gamma mc^2 \qquad (6.9)$$

for a rest mass m having velocity v, $\gamma = 1/\sqrt{1 - v^2/c^2}$, in a certain inertial frame. This definition then uses the concept of relativistic mass $m(v)$ that has proved so convenient and fundamental in our study of the inelastic collision. The fact that the relativistic mass is conserved and this conservation law satisfies Postulate 1 (it retains its form and is true in all inertial frames) means that the relativistic energy is a conserved quantity. This is the relativistic generalization of the conservation of rest mass that is an essential aspect of Newtonian mechanics.

Eq. (6.9) also suggests a definition of relativistic kinetic energy that generalizes the Newtonian quantity, $mv^2/2$. Because E reduces to mc^2 for a body at rest, the difference, $E - mc^2$, is a relativistic measure of the energy due to velocity. Call this difference the relativistic kinetic energy T,

$$T \equiv E - mc^2 = (\gamma - 1)mc^2.$$

When $v^2/c^2 \ll 1$, T reduces to the familiar quantity $mv^2/2$, because

$$T = (\gamma - 1)mc^2 \approx \left[\left(1 + \frac{1}{2}\frac{v^2}{c^2}\right) - 1\right]mc^2 = \frac{1}{2}mv^2.$$

T proves to be a handy quantity in relativistic kinematics problems.

Most textbooks on relativity work with the relativistic energy $E = \gamma mc^2$ and rest mass m rather than with the relativistic mass, a velocity-dependent quantity, $m(v) = \gamma m$. We shall do the same from here on to avoid possible confusion. Mass means rest mass m, just as in Newton's world. Relativistic energy is $E = \gamma mc^2$ and relativistic momentum is $\mathbf{p} = \gamma m\mathbf{v}$. The factor of γ will be written out explicitly.

6.3 Relativistic Force and Energy Conservation

To do problems in relativity with a given applied force that generates acceleration, we need the relativistic generalization of Newton's Second Law, $\mathbf{F} = (d/dt)\mathbf{p}$, $\mathbf{p} = m\mathbf{v}$. In many situations, $\mathbf{F}$ varies from point to point in a simple way. For example, to describe gravitational attraction between masses m and M a distance r apart, we know from Newton's law of gravity that

$$\mathbf{F} = -\frac{GmM}{r^2}\hat{\mathbf{r}},$$

where $\hat{\mathbf{r}}$ is a unit vector in the direction pointing from particle M to m, G is Newton's constant, and $\mathbf{F}$ is the force that M exerts on m. In the case of a harmonic oscillator,

$$\mathbf{F} = -kr\hat{\mathbf{r}},$$

where a particle is attracted to the origin by a force proportional to its distance away from the origin. If the particle is constrained to move along the x axis, then the harmonic force reads $-kx$. In all these cases we can introduce the concept of potential energy and write the force as the spatial rate of change of the potential energy. Because the problems we deal with here are essentially one dimensional, we illustrate this just for motion along the x axis.

In the case of the harmonic oscillator, then, it proves useful to introduce the potential energy $U(x) = (1/2)kx^2$ so

$$F(x) = -kx = -\frac{d}{dx}U(x).$$

The harmonic oscillator illustrates a general strategy: Given a general function U(x), we can calculate the force by computing the spatial rate of change $F(x) = -dU/dx$. We introduce the potential energy because it simplifies problem solving and leads to energy conservation. For example, the work that the force does on a particle of mass m when it is moved from point 1 to point 2 along the x axis is

$$W_{12} = \int_1^2 F(x)dx = -\int_1^2 \frac{dU(x)}{dx}dx = U(1) - U(2). \qquad (6.10)$$

But, using the equation of motion, we can calculate W_{12} in another way,

$$W_{12} = \int_1^2 Fdx = \int_1^2 \left(\frac{d}{dt}\right)p\,dx = \int_1^2 m\frac{d}{dt}v \cdot v\,dt,$$

where we noted that distance is velocity times time, $dx = vdt$. But the integral can be done exactly, noting that $(d/dt)(mv^2/2) = mvdv/dt$,

$$W_{12} = \int_1^2 \frac{d}{dt}\left(\frac{1}{2}mv^2\right)dt = T_2^{NR} - T_1^{NR}, \qquad (6.11)$$

where T^{NR} (the nonrelativistic kinetic energy) $= (1/2)mv^2$. Combining Eqs. (6.10) and (6.11), we have

$$U(1) - U(2) = T_2^{NR} - T_1^{NR}$$

or

$$T_1^{NR} + U_1 = T_2^{NR} + U_2.$$

In other words, the total energy $E = T^{NR} + U$ is conserved in Newton's world whenever an x-dependent static potential energy $U(x)$ exists.

In summary, using Newton's Second Law, $\mathbf{F} = d\mathbf{p}/dt$, we derive energy conservation under appropriate conditions. These two concepts are central in the solution of nonrelativistic mechanics problems.

The issue here is the generalization of these ideas to the relativistic world of Einstein. Certainly the momentum in Newton's Second Law must become its relativistic cousin,

$$\mathbf{F} = \frac{d}{dt}(\gamma m\mathbf{v}). \qquad (6.12)$$

In a force-free environment, this equation predicts conservation of the relativistic momentum, as desired. If several particles are interacting among themselves, but the total of the forces sum to zero, then the total relativistic momentum is conserved. In this way, Eq. (6.12) becomes the underpinning of our previous discussion of the inelastic collision.

Let us do a simple practice problem with Eq. (6.12) before continuing [1]. Place a charged particle of mass m in a constant electric field. The electric force will accelerate the particle and, in a Newtonian description, the particle's velocity will increase without bound. But in relativity there is a speed limit, so v will increase arbitrarily closely to c, never attaining it because of the factor of $\gamma = 1/\sqrt{1 - v^2/c^2}$ in Eq. (6.12), which grows without bound as v approaches c. In more detail, let the particle have an initial velocity of 0, $v(t = 0) = 0$, and let it be subjected to a constant force F in the x direction. Then,

$$F = \frac{d}{dt}p.$$

So, p will increase linearly with t,

$$\gamma m v = Ft. \tag{6.13}$$

Because $\gamma = 1/\sqrt{1 - v^2(t)/c^2}$, we can solve Eq. (6.13) for $v(t)$ and find, after some algebra,

$$v(t) = \frac{c}{\sqrt{1 + (mc/Ft)^2}}. \tag{6.14}$$

We see, as expected, that $v(t)$ is always less than c, but approaches it when $Ft \gg mc$. At the other extreme, if $Ft \ll mc$, then $v(t) \approx (c/(mc/Ft)) = (F/m)t \ll c$ and we retrieve Newton's result that the particle experiences a constant acceleration F/m that produces a velocity that grows linearly with time.

In summary, our relativistic dynamics has passed an important test— we cannot accelerate a massive body beyond the speed limit. Furthermore, Eq. (6.14) predicts $v(t)$ for a particle in a linear accelerator, so its functional form can be compared successfully to experimental data. Of course, in the real world there are complications, such as the fact that accelerated charged particles radiate energy, which must be accounted for in quantitative tests. Needless to say, relativistic dynamics with all the trimmings is an unparalleled success.

Our next big task is to find the relativistic generalization of energy conservation for forces that are obtained from a static potential. As in the non-relativistic world,

$$W_{12} = \int_1^2 F dx = -\int_1^2 \frac{dU}{dx} dx = U(1) - U(2).$$

Now we need to use the equation of motion to see how the force changes the particle's kinetic energy:

$$W_{12} = \int_1^2 F dx = \int_1^2 \frac{d}{dt}(\gamma m v) v \, dt. \tag{6.15}$$

Can we write the integrand as a total time derivative? We hope that the answer will involve the relativistic energy,

$$E = \gamma m c^2.$$

So, consider

$$\frac{d}{dt}(\gamma m c^2) = \frac{d}{dt}\left(\frac{mc^2}{\sqrt{1 - v^2/c^2}}\right) = -\frac{1}{2}\gamma^3 m c^2 \cdot \left(-\frac{2v}{c^2}\frac{dv}{dt}\right) = \gamma^3 m v \frac{dv}{dt}.$$

But the integrand of Eq. (6.15) reads

$$\frac{d}{dt}(\gamma m v) = \gamma m \frac{dv}{dt} - \frac{1}{2}\gamma^3 m v \cdot \left(-\frac{2v}{c^2}\frac{dv}{dt}\right) = \gamma m \left(1 + \gamma^2 \frac{v^2}{c^2}\right)\frac{dv}{dt}.$$

Finally,

$$\frac{d}{dt}(\gamma m v) = \gamma m \left(1 + \frac{v^2/c^2}{1 - v^2/c^2}\right)\frac{dv}{dt} = \gamma^3 m \frac{dv}{dt},$$

which we identify as the time rate of change of the relativistic energy calculated before. So, things have worked out perfectly:

$$W_{12} = \int_1^2 \frac{d}{dt}(\gamma m c^2) dt = E_2 - E_1 = T_2 - T_1. \tag{6.16}$$

Combining Eqs. (6.15) and (6.16),

$$T_1 + U_1 = T_2 + U_2, \tag{6.17}$$

and the total energy, relativistic kinetic and potential, is conserved under these conditions. Note that since $T = E - mc^2$, we could put either T or E into Eq. (6.17), and we used the relativistic kinetic energy just to mirror the nonrelativistic discussion.

6.4   Energy and Momentum Conservation, and Four-Vectors

Our formulas for the relativistic momentum and energy, $\mathbf{p} = \gamma m \mathbf{v}$ and $E = \gamma m c^2$, are much simpler than their derivations. We aim to remedy this problem here. Both of these quantities were determined to satisfy Postulate 1. The factors of γ are forced on us by the transformation laws of space and time measurements between inertial frames. In Newton's world, the momentum inherits its transformation properties from those of space and time,

$$\mathbf{p} = m\frac{d\mathbf{x}}{dt} \qquad \text{(Newton)}. \qquad (6.18)$$

$\mathbf{x}$ transforms according to Galileo, $x = x' + vt$, $y' = y$, $z' = z$, and t is universal, $t = t'$. Equation (6.18) then implies that $\mathbf{p}$ transforms as a velocity, and because velocities add in Newton's world, we can show that if there is momentum conservation in one inertial frame, there is momentum conservation in all.

Can we write a relativistic generalization of Eq. (6.18)? The key in Eq. (6.18) is that the numerator is a distance that transforms simply and the denominator is an invariant. Because the proper time τ is an invariant in a relativistic world, we should try

$$\mathbf{p} = m\frac{d\mathbf{x}}{d\tau} \qquad \text{(Einstein)}. \qquad (6.19)$$

Eq. (6.19) reproduces $\mathbf{p} = \gamma m\mathbf{v}$ because $d\tau$ is related to the time interval dt by time dilation, $d\tau = \sqrt{1 - v^2/c^2}\, dt$. The truly crucial feature of Eq. (6.19) is that $\mathbf{p}$ transforms between inertial frames such as $\mathbf{x}$ because $d\tau$ is an invariant.

There is still a puzzle here. When $\mathbf{x}$ transforms between frames, the time variable mixes in. In other words, it takes *four* quantities, the three components of $\mathbf{x}$ and t, to write an expression for the transformation law of any one of them. So, Eq. (6.19) must be supplemented by an expression involving t. Clearly, $dt/d\tau$ is the first candidate to come to mind and because this derivative is γ, we are led to

$$E = mc^2\frac{dt}{d\tau} = \gamma mc^2,$$

where hindsight led us to include the factor of c^2 and identify the relativistic energy. So, E gives the zeroth component of a four-vector of energy-momentum. We write

$$p_\mu = \left(\frac{E}{c}, p_1, p_2, p_3\right),$$

where μ is an index that can take the values 0, 1, 2, or 3 and $p_o = E/c$, and so on. Our original four-vector is space-time,

$$x_\mu = (ct, x_1, x_2, x_3),$$

and the crucial feature about this quantity is its transformation law under boosts, the Lorentz transformation,

$$x_1' = \gamma(x_1 - vt)$$

$$x_2' = x_2$$

$$x_3' = x_3$$

$$t' = \gamma\left(t - \frac{v}{c^2}x_1\right).$$

The transformation laws for $\mathbf{p}$, E' are now immediate:

$$p_1' = m\frac{dx_1'}{d\tau} = \gamma m\left(\frac{dx_1}{d\tau} - v\frac{dt}{d\tau}\right) = \gamma\left(p_1 - \frac{v}{c^2}E\right)$$

$$p_2' = p_2$$

$$p_3' = p_3 \tag{6.20}$$

$$E' = mc^2\frac{dt'}{d\tau} = \gamma mc^2\left(\frac{dt}{d\tau} - \frac{v}{c^2}\frac{dx_1}{d\tau}\right) = \gamma(E - vp_1).$$

Knowing the transformation law Eq. (6.20) will help us solve problems in relativistic collisions.

The reader should be careful not to confuse the γ factors in the expression for $\mathbf{p}$ and E with the γ factors in Eq. (6.20). In the first case, γ contains the velocity of the particle in a given frame, and in the transformation formulas Eq. (6.20), γ contains the relative velocity between two inertial frames.

We learn several points from Eq. (6.20). First, momentum and energy conservation are unified. Because momentum and energy mix under a boost, we must have conservation of all four components of the energy-momentum four-vector together. It would be inconsistent with Postulate 1 to have fewer. Second, because $\mathbf{p}$ and E form a four-vector, we should be able to construct an invariant frame-independent quadratic form in analogy to the invariant interval

$$c^2t^2 - x_1^2 - x_2^2 - x_3^2 = c^2t'^2 - x_1'^2 - x_2'^2 - x_3'^2.$$

For energy-momentum, we have

$$\frac{1}{c^2}E^2 - \mathbf{p}^2 = \frac{1}{c^2}\gamma^2 m^2 c^4 - \gamma^2 m^2 v^2 = \gamma^2 m^2(c^2 - v^2)$$

$$= \frac{m^2 c^2}{1 - v^2/c^2}(1 - v^2/c^2) = m^2 c^2.$$

So, we have derived the energy-momentum relation for a particle of mass m,

$$E^2 = \mathbf{p}^2 c^2 + m^2 c^4. \tag{6.21}$$

This is the relativistic version of the energy-momentum relation of Newtonian physics,

$$T^{NR} = \frac{1}{2}mv^2 = \frac{\mathbf{p}^2}{2m}. \tag{6.22}$$

Note that Eq. (6.21) is quite different from Eq. (6.22). Substituting the relativistic kinetic energy $T = E - mc^2$ into Eq. (6.21) gives

$$T^2 + 2mc^2 T = \mathbf{p}^2 c^2. \tag{6.23}$$

If $v/c \ll 1$, then we observed earlier that $T \approx (1/2)mv^2$ so $T \ll mc^2$ and Eq. (6.23) reduces to

$$T \approx \mathbf{p}^2/2m$$

as expected. But for $v \approx c$, $pc \gg mc$, so Eq. (6.21) becomes

$$E \approx pc.$$

Therefore, E and p, the magnitude of $\mathbf{p}$ ($p \equiv \sqrt{p_1^2 + p_2^2 + p_2^2}$), become linearly related.

Finally, note that if the rest mass of the particle vanishes, it can still carry momentum and energy, and we have exactly

$$E = pc. \tag{6.24}$$

This is the exact energy-momentum relation for light, as we mentioned early in this chapter. Light travels at the speed limit because it has no rest mass. Maxwell's wave theory of light predicts Eq. (6.24) from first principles. This relation is particularly important in the quantum theory of light, as we illustrate next.

6.5 Collisions and Conservation Laws—Converting Mass to Energy and Energy to Mass

It is interesting to consider relativistic collisions and decay processes that illustrate the energy-momentum conservation laws. Some of these processes involve light that satisfies the energy-momentum relation, $E = pc$. In addition, if we deal with individual quanta of light, then the energy comes in a packet $E = h\nu$, where ν is the frequency of the light wave and h is Planck's constant, $h = 6.627 \cdot 10^{-34}$ J-s, which sets the scale of quantum physics. We certainly won't be doing any quantum mechanics here, so we just borrow $E = h\nu$ to illustrate relativity in interesting settings involving elementary particles.

As our first example, consider a nucleus of rest mass M_o that absorbs a photon of energy Q. Our final state consists of an excited nucleus of rest

Before   After

FIGURE 6.9 ▶

mass M' and recoil velocity v. We want to find M' and v in terms of M_o and Q. The process is shown in Figure 6.9. Energy conservation reads

$$Q + M_o c^2 = \gamma M' c^2$$

and momentum conservation reads

$$Q/c = \gamma M' v.$$

Combining these equations, we find

$$\frac{v}{c} = \frac{Q}{\gamma M' c^2} = \frac{Q}{Q + M_o c^2},$$

which gives the recoil velocity. Finally, we can solve for M',

$$M' = M_o \sqrt{1 + 2Q/M_o c^2}. \tag{6.25}$$

Note that $M' c^2$ is less than the sum of the initial relativistic energies, $M_o c^2 + Q$, because some energy goes into M''s recoil. This is how the kinematics works out. However, it might be that the process is not possible and does not occur. The point is that the laws of nuclear physics (quantum mechanics) predict a discrete list of possible states and possible M' values for each nucleus. If the M' of Eq. (6.25) doesn't match one of these allowed values, the initial photon will not be absorbed. Scattering of the photon might be the actual physical event in that case.

It is interesting to turn this process around and consider photon emission from an excited atom. The process is shown in Figure 6.10. Conservation of energy and momentum reads

$$M_o c^2 = E' + Q$$

$$0 = p' - Q/c, \tag{6.26}$$

M_o

$v \longleftarrow M'_o \quad \rightsquigarrow Q$

Before   After

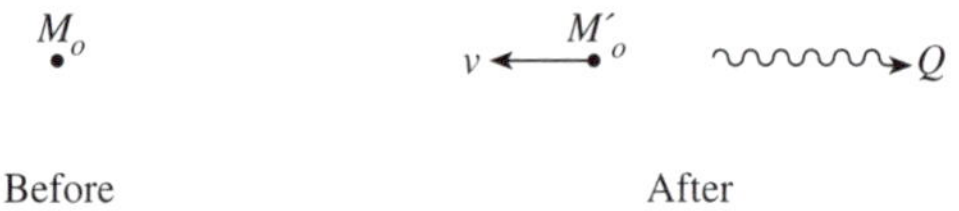

FIGURE 6.10 ▶

where E' and p' label the relativistic energy and momentum of the final (unexcited) atom of mass M'_o. Let's say that we detect the photon and measure its energy Q. So, we need to eliminate E' and p' from Eq. (6.26). A slick way to do this is to use the energy-momentum relation $E'^2 = p'^2 c^2 + M'^2_o c^4$. Re-arranging Eq. (6.26) gives

$$E' = M_o c^2 - Q$$

$$p' = Q/c.$$

So,

$$E'^2 - p'^2 c^2 = (M_o c^2 - Q)^2 - Q^2 = M'^2_o c^4.$$

Finally,

$$M_o^2 c^4 - 2M_o c^2 Q = M'^2_o c^4. \qquad (6.27)$$

Let's write this in terms of the energy difference between the initial and final atoms, taken at rest,

$$\Delta E \equiv M_o c^2 - M'_o c^2. \qquad (6.28)$$

We focus on ΔE because this would be the energy difference between discrete quantum energy levels of the atom. One could calculate such a difference from first principles in quantum mechanics. Solving Eq. (6.28) for $M'_o c^2$,

$$M_o c^2 - \Delta E = M'_o c^2.$$

Squaring gives

$$M_o^2 c^4 - 2M_o c^2 \Delta E + (\Delta E)^2 = M'^2_o c^4.$$

Combining this with Eq. (6.27) gives

$$Q = \Delta E \left(1 - \frac{\Delta E}{2M_o c^2}\right). \qquad (6.29)$$

This is the desired result. We learn that recoil, a consequence of energy-momentum conservation, has reduced the energy of the photon from ΔE to $\Delta E(1 - \Delta E/2M_o c^2)$. If we observe the photon and measure Q, we must use Eq. (6.29) to predict ΔE to compare with calculational predictions of quantum mechanics. For heavy atoms, the recoil is a small effect. For example, consider ^{198}Hg, which emits photons with an energy 412 keV. Because $M_o = 198$ a.m.u. $= 3.28 \cdot 10^{-25}$ kg, we compute $\Delta E/2M_o c^2 \approx 10^{-6}$. This is small, but not really negligible in our quantum world. For example, if the emitted photon were incident on another ^{198}Hg, it could not be absorbed because its energy does not quite match the needed ΔE—it is not quite energetic enough to cause the transition.

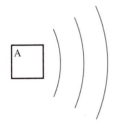

FIGURE 6.11 ▶

There is a way around this dilemma, first developed by Mössbauer. Let the atom that emits the photon be part of a large regular crystal. Then, when it emits the photon, the recoiling momentum, if it is sufficiently small, is imparted to the entire crystal! The recoil term, $\Delta E/2M_o c^2$, is now reduced by the number of atoms in the crystal and becomes truly negligible, less than the intrinsic energy spread of each spectral line. Using this observation, Mössbauer studied spectral lines in detail. His experimental apparatus considered two crystals, one emitting and the other absorbing, at a relative velocity v, as in Figure 6.11. If v is too large, then the energy of the photons in the rest frame of the atoms in crystal B is too large to match the energy difference between the quantum states, even accounting for their intrinsic uncertainties (widths). In fact, a tiny v on the order of a few centimeters per second, is sufficient to produce a photon energy that misses ΔE. By varying v, the detailed structure of spectral lines can be mapped out.

Now let's turn to several illustrations of particle creation in high energy physics collisions. At an accelerator center, researchers convert kinetic energy into mass and search for new elementary particles. Colliding beams are a particularly effective way of achieving this goal, and important discoveries of new resonances, evidence for the existence of heavy quarks, were made this way. Consider proton–proton collisions in the center of momentum frame and suppose we wish to create a pion, the particle responsible for the nuclear force as you will learn in quantum mechanics:

$$P + P \to P + N + \pi^+.$$

The collision is shown in Figure 6.12. We want to know the minimal velocity u that will make it possible to create the extra particle, a pion. The rest

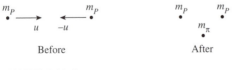

Before   After

FIGURE 6.12 ▶

masses involved are

$$m_\pi/m_P \approx 0.149, \qquad m_P c^2 \approx 938 \text{ MeV}$$

and the proton and neutron have approximately equal masses (actually the neutron is slightly heavier than the proton, approximately 0.1%). We can compute the required u, and thus know how powerful an accelerator is required. Using relativistic energy conservation,

$$2\gamma m_P c^2 = 2 m_P c^2 + m_\pi c^2.$$

Therefore,

$$\gamma = \frac{1}{\sqrt{1 - u^2/c^2}} = 1 + \frac{m_\pi}{2 m_P} \approx 1.074.$$

Doing the arithmetic gives

$$u/c \approx 0.37.$$

To put this result into perspective, imagine a collision at another accelerator center where one proton is speeding along in a beam and the other proton is in a stationary target. How fast u would the proton in the beam have to be in order to create a pion again? We can get this answer by boosting our center of momentum analysis. Consider a frame S moving to the left in Figure 6.12 at velocity u, so that the proton with velocity $-u$ in the center of momentum frame is brought to rest (Figure 6.13). The left proton has a velocity

$$v = \frac{u + u}{1 + u^2/c^2}$$

in the lab frame, using the addition of velocity formula. Since $u/c = 0.37$, we learn that $v = 0.65c$ and the relativistic kinetic energy of the beam proton is $(\gamma - 1)mc^2 = (1/\sqrt{1 - v^2/c^2} - 1)mc^2 \approx (1.31 - 1)$ 938 MeV $= 290$ MeV. We learn from this that the proton in the beam needs to have more than twice the kinetic energy of the rest mass of the pion we are trying to create. Mass creation in this sort of experiment is much less efficient than that in the center of momentum frame because of momentum conservation—the three

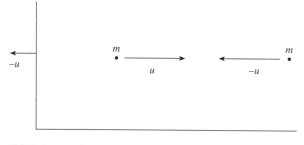

FIGURE 6.13 ▶

particles in the final state must have net momentum to the right to match the momentum of the initial proton in the beam, so we cannot convert all the kinetic energy to mass. This calculation illustrates why colliding beam experiments are best in exploring the world of new, higher energy states of matter. Unfortunately, it is harder to make colliding beam machines with beams intense enough to make such experiments practical than it is to make fixed target accelerators.

Finally, let us consider a scattering process in three dimensions. Consider Compton scattering, the elastic scattering of a photon off an atom or any charged particle. Studies of this process were important in establishing the quantum theory of light. A photon of energy Q_o scatters off a stationary electron and the photon emerges at an angle θ with diminished energy Q, while the electron recoils through angle φ with final energy E and momentum $\mathbf{p}$. This is shown in Figure 6.14, where we have labelled the direction of the initial photon with the unit vector $\hat{\mathbf{n}}_o$ and the direction of the final photon with the unit vector $\hat{\mathbf{n}}$. Writing out the conservation laws,

$$Q_o + mc^2 = E + Q$$
$$\hat{\mathbf{n}}_o Q_o / c = \hat{\mathbf{n}} Q / c + \mathbf{p}.$$

$$(6.30)$$

Suppose our experiment just detects the final photon, so we want to eliminate E and φ from the kinematics. Solving Eq. (6.30) for E and $\mathbf{p}c$ gives

$$E = (Q_o - Q) + mc^2$$
$$\mathbf{p}c = (\hat{\mathbf{n}}_o Q_o - \hat{\mathbf{n}} Q).$$

Squaring each equation gives

$$(Q_o - Q)^2 + 2(Q_o - Q)mc^2 + m^2 c^4 = E^2$$
$$Q_o^2 - 2Q_o Q \cos\theta + Q^2 = c^2 \mathbf{p}^2$$

where we identified $\hat{\mathbf{n}} \cdot \hat{\mathbf{n}}_o = \cos\theta$. Now subtract these equations using the energy-momentum relative for the recoiling electron, $E^2 = \mathbf{p}^2 c^2 + m^2 c^4$:

$$2Q_o Q(1 - \cos\theta) - 2(Q_o - Q)mc^2 = 0.$$

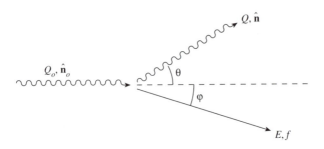

FIGURE 6.14 ▶

Dividing through by $2Q_oQmc^2$ gives

$$\frac{1}{Q} = \frac{1}{Q_o} + \frac{1}{mc^2}(1 - \cos\theta). \tag{6.31}$$

So, if the photon scatters through an angle θ, its energy is diminished from Q_o to Q according to this, the Compton formula. This result is usually quoted in the context of quantum mechanics where the energy of a single quantum of light is related to its frequency ν by Planck's constant h, $h = 6.627 \cdot 10^{-34}$ J-s,

$$Q = h\nu.$$

Because $\nu = c/\lambda$ for a wave traveling with the speed of light, we can replace $1/Q$ by λ/hc in Eq. (6.31) and get

$$\lambda - \lambda_o = \frac{h}{mc}(1 - \cos\theta).$$

In other words, when light scatters, its wavelength increases proportionally to 1 minus the cosine of its scattering angle. Setting the scale of the effect with Planck's constant, $h = 6.627 \cdot 10^{-34}$ J-s, the characteristic length in the Compton formula,

$$\frac{h}{mc} \approx 0.02426 \text{ Å} = 2.4 \cdot 10^{-10} \text{ m},$$

which is often referred to as the Compton wavelength of the electron, a very small distance on the scale of atomic physics. If the initial photon were an X-ray with $\lambda_o = 0.7$Å (angstrom, Å, is a convenient unit of length in atomic physics, $\text{Å} \equiv 10^{-10}$ m) and we take scattering through 45°, then

$$\lambda = \lambda_o + \frac{h}{mc}(1 - \cos\theta) = 0.7 + 0.02426(0.7071) \approx 0.7018 \text{ Å}.$$

Scattering experiments such as this, coupled with quantum mechanics calculations of the cross sections involved, played a central role in the development of atomic physics.

6.6   Further Topics in Special Relativity

This completes our exposition of special relativity. It is a pity to stop here because the best is yet to come. As you continue your physics studies, you will encounter many other topics where special relativity plays a central role. The most insightful of all these are electricity and magnetism. Perhaps you know about static charges and electric fields, on the one hand, and currents and magnetic fields, on the other. Special relativity unifies these ideas into one whole that elucidates both parts. For example, if there is a static charge and attendant electric field in a reference frame S, then in the frame S', which moves at a relative velocity v, you will observe a current as well which produces a magnetic field. This means that electric and magnetic fields must transform into one another under boosts—it is inconsistent to think of electric effects without magnetic effects and vice versa! In the frame S, a charged particle experiences electric forces, but when viewed from frame S′ where there are magnetic fields, the charged particle also experiences magnetic forces that act perpendicular to the particle's velocity. All these effects, for example, the Lorentz force law, with all the details attached, can be derived from scratch from static electric effects using special relativity. It is a beautiful subject that you will learn about in your next course on electricity and magnetism.

The applications of special relativity to more practical subjects are also important. Relativistic plasmas, hot gases of electrons confined to magnetic bottles, and their properties are central in taming the energy source of the Sun, fusion, and making fusion reactors for energy production. Relativistic particle beam physics, with electrons, protons, and more exotic elementary particles moving within a fraction of a percent of the speed limit, is an important subject in designing new accelerators. In fact, accelerators are no longer the sole domain of high energy experiments. More and more applications of accelerators to medical treatments, imaging, and structural studies of materials are being developed at an impressive pace. Your understanding of special relativity will serve you well in the future.

▶ Problems ▶

6-1. Our galaxy is approximately 10^5 light-years across, and the most energetic naturally occuring particles have an energy of approximately 10^{19} eV. How long would it take a proton with this energy to travel across the galaxy as measured in the rest frame of (a) the galaxy or (b) the particle?

6-2. An electron is accelerated from rest through a voltage drop of 10^5 volts and then travels at constant velocity.

   (a) How long does it take the electron to travel 10 m, after it has reached its final velocity?

   (b) What is the distance as measured in the rest frame of the electron?

6-3. If all the energy used in running the accelerator at Fermilab for a full day could be collected in a box, how much heavier would the box become. (Find out about the energy requirements at Fermilab through their Website, www.fnal.gov.)

6-4. Plot the total energy E versus momentum p for a particle of rest mass m_o for the cases (a) Newtonian kinematics, (b) relativistic kinematics, and (c) relativistic kinematics for a massless particle, $m_o = 0$. Note the ranges in momentum where two or more of the three curves are good approximations of one another.

6-5. In the chapter, we derived the Lorentz transformation law for energy and momentum between a frame S and a frame S' moving to the right at velocity v. Specialize to the case of the energy and momentum carried by light, $P = E/c$, and show that the transformation for radiant energy between frames S and S' for light traveling in the x direction is

$$E' = \sqrt{\frac{1 - v/c}{1 + v/c}} E.$$

Note that this result is algebraically the same as the Doppler shift for the frequencies of light observed in the two frames. In other words, the transformation laws for energy and frequency of light are identical. This is an important result for both the basic theory of electromagnetic phenomena and quantum mechanics, in which the quantum of electromagnetic energy, the photon, carries energy proportional to its frequency, $E = h\nu$, where h is Planck's constant, the cornerstone of quantum theory.

6-6. Radiant energy from the Sun is received on Earth at a rate of about 1370 J/s-m^2 on a surface perpendicular to the Sun's rays.

(a) What total force would be exerted on all of Earth if all the light energy were absorbed?

(b) Is this a large or a small force? (Compare it to the total gravitational force that the Sun exerts on the Earth.)

6-7. A star of mass 10^{32} kg is surrounded by a thin, flexible spherical shell of mass 10^{25} kg. The star loses mass at the rate 10^{10} kg/s in the form of light. Suppose that all of this radiant energy is absorbed on the shell. What must the radius of the shell be so that the radiation pressure from the light balances its gravitational attraction to the star?

6-8. Light rays from the Sun hit the Earth at the rate of 1370 J/s-m^2 on a surface perpendicular to the Sun's rays.

(a) How much mass in the Sun is converted to energy per second to account for the radiant energy hitting Earth? (The radius of Earth is roughly $6.4 \cdot 10^6$ m and the distance from Earth to the Sun is approximately $1.5 \cdot 10^{11}$ m)

(b) What is the total mass converted to energy in the Sun to supply this radiant energy?

(c) Estimate the mass of hydrogen that must be converted to helium per second to supply this radiant energy. (Recall that most of the Sun's energy comes from fusing hydrogen into helium. The mass of a hydrogen nucleus is $1.67262 \cdot 10^{-27}$ kg and the mass of a helium nucleus is $6.64648 \cdot 10^{-27}$ kg.)

(d) Estimate how long the Sun will warm Earth, accounting only for the hydrogen fusion process.

6-9. The physicist Sir Arthur Eddington, who did pioneering work in cos-
mology, pointed out the strength of unscreened electrostatic forces in
the following dramatic fashion. Take 1 g of electrons and place them
uniformly at rest in a spherical container of radius 10 cm. Calculate the
mass associated with their electrostatic potential energy and verify that it
is on the order of 10^{10} kg! (Recall from electricity and magnetism that
electrons repel one another through the inverse square law $F = kq^2/r^2$,
where q is the electronic charge in coulombs and k is approximately
$9 \cdot 10^9$ newton-m^2/coulomb^2. The electrostatic energy of a uniformly
charged sphere of total charge Q and radius r is $3kQ^2/5r$. The charge
of an electron is $q = 1.6 \cdot 10^{-19}$ coulombs, and its mass is $m = 9.11 \cdot 10^{-31}$ kg.)

This alarming answer shows how important the neutrality of bulk matter
really is. The negative charges of the electrons in the paper of this book,
which weighs under a pound, are neutralized by the positive charges of
the protons in the nuclei of the atoms that bind them, and the powerful
long-range electromagnetic forces that interested Sir Arthur Eddington
are canceled out.

6-10. Our derivation of $E = mc^2$ in the chapter using Einstein's box made two
simplifications. First, we ignored the distance that the box recoils when
light is in transit from one end to the other. And second, we ignored the
decrease in the mass of the box when the light is in transit. Include these
effects in a calculation of the same problem and show that $E = mc^2$ can
be derived perfectly.

6-11. A particle is given a relativistic kinetic energy equal to twice its rest mass.
Find its resulting speed and momentum. How do these results change if
the relativistic kinetic energy is five times the particle's rest mass?

6-12. Through what voltage would you have to accelerate an electron in order
to boost its speed from rest to 99% of the speed of light? Repeat your
calculation for a proton.

6-13. A relativistic electron moving in the x direction enters a region of space
where there is a uniform electric field in the y direction.

(a) Write down the relativistic equations of motion that describe this situation. Following Section 6.3, you will have two equations, one for
the x component of the force and one for the y component.

(b) Discuss the solution of the equations in part (a) qualitatively. For example, identify conserved quantities and time variable quantities. Show that
the x component of the velocity of the particle decreases with time.

(c) Solve for $v_x(t)$ and $v_y(t)$, assuming that initially the electron had a
velocity v_o in the x direction. Denote the component of the force in
the y direction F_y.

(d) Suggest some practical uses for a device based on your results in an
accelerator center where the electrons have v_o values very close to
the speed limit c.

6-14. In science fiction stories, we find space vehicles that consist of large sails that deflect light so that the recoil of the light propels the vehicle to amazing speeds. Suppose that the Department of Energy has funded such a project and there is a sail-ship in free space that feels the push of a strong, steady laser beam of light directed at it from Earth. If the sail is perfectly reflecting, calculate the mass equivalent of light required to accelerate a vehicle of rest mass m_o up to a fixed γ.

6-15. A laser with a mass of 10 kg is in free space with its beam directed toward Earth. The laser continuously emits 10^{20} photons/s of wavelength 6000 Å, as measured in its own rest frame. At $t = 0$ the laser is at rest with respect to Earth.

(a) Initially how much radiant energy per second is received on Earth? (Planck's constant $h = 6.627 \cdot 10^{-31}$ kg-m^2/s, and the energy of a photon is $E = h\nu$.)

(b) The radiation emitted toward Earth causes the laser to recoil away from Earth. What is the velocity of the laser relative to Earth after 10 years have elapsed on a clock at rest with respect to the laser?

(c) At the time when the laser is moving with velocity v relative to Earth, how much less is the rate at which energy is received on Earth than the original rate when $v = 0$? Evaluate this for $t = 10$ years, laser time.

(d) Show how an observer on Earth can explain the continually decreasing rate of reception in terms of energy considerations.

6-16. A photon rocket uses light as a propellent. If the initial and final rest masses of the rocket are M_i and M_f, show that the final velocity v of the rocket relative to its initial rest frame is given by the equation

$$M_i/M_f = \sqrt{(1+v/c)/(1-v/c)}.$$

▶ **Problems on Collisions and Conservation Laws** ▶ __

6-17. When a K^o meson decays at rest into a π^+ and a π^- meson, each escapes with a speed of approximately $0.85c$. Now consider a K^o meson that is traveling at a speed of $0.9c$ relative to the lab frame. If it then decays, what is the greatest speed that one of the pions can have in the lab frame? What is the least speed?

6-18. Two identical particles, A and B, are approaching each other along a common straight line. Each particle has the same speed v, as measured in the lab. Show that the energy of particle A as measured by B is $(1 + v^2/c^2)(1 - v^2/c^2)^{-1}M_o c^2$, where M_o is A's rest mass.

6-19 A photon has energy 200 MeV and is traveling along the x axis. Suppose another photon has energy 400 MeV and is traveling along the y axis.

(a) What is the total energy of this system? What is its total momentum?

(b) If a single particle had the same energy and momentum, what would its mass be? What would its direction of travel and what would its speed be?

6-20. A particle of rest mass m_o and relativistic kinetic energy $3m_oc^2$ strikes and sticks to a stationary particle of rest mass $2m_o$. Find the rest mass M_o of the resulting composite particle.

6-21 (a) A photon of energy E is absorbed by a stationary particle of rest mass m_o. What is the velocity and rest mass of the resulting composite particle?

(b) Repeat part (a) replacing the photon by a particle of rest mass m_o and speed $0.8c$.

6-22. An atom in an excited state of energy Q_o above the ground state moves toward a scintillation counter with speed v. The atom decays to its ground state by emitting a photon of energy Q, as measured by the counter, coming completely to rest as it does so. If the rest mass of the atom is m, show that $Q = Q_o[1 + (Q_o/2mc^2)]$.

6-23. A neutral pion decays into two photons. The pion's rest mass is 135 MeV. Suppose it is in a secondary beam at Fermilab with a relativistic kinetic energy of 1 GeV.

(a) What are the energies of the photons if they are emitted in opposite directions along the pion's original line of motion.

(b) What angle is formed between the two photons if they are emitted at equal angles to the direction of the pion's motion?

6-24. An anti-proton of kinetic energy 1 GeV strikes a proton at rest in the lab. (The proton and its anti-particle have identical masses, about 938 MeV/c².) They annihilate, and two photons emerge from the reaction, one traveling forward and one backward along the beam direction.

(a) What are the energies of the two photons?

(b) As measured in a reference frame of the anti-proton, what energy does each photon have?

6-25. According to Newtonian mechanics, when a beam particle collides off an identical particle originally at rest, they emerge with an angle between them of exactly 90° in all cases. Contrast this result to a relativistic collision:

(a) If a proton of kinetic energy 500 MeV collides elastically with a proton at rest, and the protons rebound with equal energies, what is the angle between them?

(b) Repeat this exercise with a proton having an initial kinetic energy of 100 GeV.

6-26. Supppose a photon has a head-on collision with an electron. What initial velocity must the electron have if the collision results in a photon recoiling straight backward with the same energy Q as it had initially?

6-27. A photon of energy E collides elastically with an electron at rest. After the collision, the photon's energy is reduced by half and its scattering angle is 60°.

(a) What was its original energy? Is the frequency of this photon in the visible range?

(b) A photon of energy E collides with an excited atom at rest. After the collision, the photon has the same energy, but its direction has changed by 180°. If the atom is in its ground state after the collision, what was its original excitation energy?

6-28. A K meson, rest mass approximately 494 MeV/c^2, decays into two pions, rest mass approximately 137 MeV/c^2. One of the pions emerges from the decay process at rest.

(a) What is the energy of the other pion?

(b) What was the energy of the original K meson?

6-29. An electron–positron pair can be produced by a gamma ray striking a stationary electron, $\gamma + e^- \rightarrow e^- + e^+ + e^-$. What is the minimum gamma ray energy that will allow this process to occur? (The positron is the electron's anti-particle. It has the same rest mass as the electron, 0.511 MeV/c^2, but the opposite charge.)

6-30. Suppose that an accelerator can give protons a kinetic energy of 300 GeV. (The rest mass of a proton is roughly 938 MeV/c^2.) Calculate the largest possible rest mass M_X of a new particle X that could be produced when the beam proton hits a stationary proton in a target, $p + p \rightarrow p + p + X$.

6-31. A positron of kinetic energy 0.511 MeV annihilates with an electron at rest, creating two photons. One photon emerges at an angle 90° to the incident positron direction.

(a) What are the energies of both photons? (The rest mass of an electron is 0.511 MeV/c^2. The rest mass of the positron is exactly the same.)

(b) What is the direction of the second photon?

6-32. A particle of kinetic energy K collides elastically with an identical particle at rest. The two outgoing particles emerge with equal and opposite angles $\theta/2$ with respect to the incoming particle. Find the energy and momentum of each outgoing particle. Find the angle θ in terms of K and the rest mass m of one of the initial particles.

6-33. A berylium nucleus consists of four protons and three neutrons. The mass of the berylium nucleus is 6536 MeV/c^2, the proton is 938.28 MeV/c^2, and the neutron is 939.57 MeV/c^2.

(a) Find the binding energy of berylium. (The binding energy of a nucleus is the difference between its rest mass and the sum of the rest masses of its free constituents.)

(b) Consider the reaction in which an extra neutron at rest is absorbed by a berylium nucleus also at rest, which subsequently decays into two alpha particles. (An alpha particle consists of two protons and two neutrons bound together. Its rest mass is 3728 MeV/c^2.) What is the kinetic energy of each of the alpha particles?

Chapter 7

A Gentle Introduction to General Relativity

7.1 | The Equivalence Principle, Gravity, and Apparent Forces

Everyone knows the story of Isaac Newton and the falling apple. There was a plague in Great Britain, so the students were sent out of the cities to reduce their chances of catching the contagion. Supposedly Newton relaxed under an apple tree, contemplating the current ideas of mechanics. When an apple fell on his head, he invented the idea of the gravitational force—Earth's enormous mass exerted a force on the apple, breaking its stem and causing a collision with Newton's precious head. This event led Newton, over the course of later months back at the university, to the force law of gravity,

$$\mathbf{F}_{12} = -G\frac{m_1 m_2}{r_{12}^2}\hat{\mathbf{r}}_{12}, \qquad (7.1.1)$$

where G is Newton's constant ($G \approx 6.67 \cdot 10^{-11}$ Nm^2/kg^2), which sets the scale for gravitational forces; m_1 and m_2 are the masses, which are a distance r_{12} apart; $\mathbf{F}_{12}$ is the force that m_1 exerts on m_2; and $\hat{\mathbf{r}}_{12}$ is the unit vector pointing from m_1 to m_2. Newton arrived at the inverse square character of the force law, Eq. (7.1.1), to explain the extensive planetary data accumulated by Kepler and others. Choosing m_1 to be the mass M_s of the Sun and m_2 to be the mass of Earth M_e, the equation of motion of the Earth

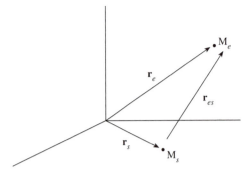

FIGURE 7.1 ▶

around the Sun is given by Newton's Second Law, force equals mass times acceleration (see also Figure 7.1).

$$M_e\ddot{\mathbf{r}}_e = -G\frac{M_eM_s}{|\mathbf{r}_e - \mathbf{r}_s|^2}\hat{\mathbf{r}}_{es}. \qquad (7.1.2)$$

A crucial element of Eq. (7.1.2) is the fact that the mass of the accelerating body, Earth in this case, cancels out of the equation of motion. We describe this by saying that the inertial mass, the mass on the left-hand side of Newton's Second Law (mass times acceleration equals force), equals the gravitational mass, the mass in the gravitational force expression. It is said that Galileo was the first physicist to investigate this point, before Newton codified classical dynamics, and establish the equality of these two masses experimentally. Galileo, through his assistants, dropped masses off the Leaning Tower of Pisa and observed that they accelerated identically in the gravitational field provided by Earth. Modern experiments by Eötvos and others have established the equality of the gravitational and inertial masses to high precision. The axiom that the two masses are strictly identical evolved into a central ingredient in the soon-to-be famous Equivalence Principle. Under Einstein, the Equivalence Principle developed into the statement that there is no way to distinguish the local effects of a gravitational field from those in an accelerating reference frame free of external masses. The Equivalence Principle allows us to understand accelerating reference frames in terms of gravity and gravity in terms of accelerating reference frames. This principle is explained and discussed in much greater detail as we journey forth.

The Equivalence Principle and the inverse square law of gravity are both under constant experimental scrutiny by high precision experiments. We accept both ideas as exact throughout our discussions. However, if one or both should fail ever so slightly, many topics of theoretical physics would need fundamental changes.

Classical physicists understood that the Principle of Equivalence made gravity a very special phenomenon. Other forces (such as electrostatics) or mechanical devices (such as springs) produce accelerations that are inversely proportional to the mass of the body. There are forces besides gravity that are familiar from day-to-day experience that produce accelerations that are independent of the mass of the body. These are called apparent forces and are strictly geometrical in origin. For example, when you drive a car and accelerate from a stop sign or decelerate at a red traffic light, you experience forces of this type. Another example consists of centripetal and Coriolis forces. These are the forces that occur when you measure acceleration in a rotating coordinate system, called a non-inertial frame of reference. Recall that an inertial frame is one in which an isolated body moves in a straight line at constant velocity. To change the body's velocity, a force must be applied. The velocity of the body changes according to Newton's Second Law, $\mathbf{f} = m\mathbf{a}$, where $\mathbf{f}$ is the applied force and $\mathbf{a}$ is the body's acceleration, $\mathbf{a} = d^2\mathbf{r}/dt^2$. A simple example of a rotating, non-inertial frame is afforded by a turntable spinning at a constant angular velocity ω in an otherwise inertial environment. From the perspective of a coordinate system fixed in the rotating turntable, a body moving at a constant velocity in the inertial frame is accelerating. Clearly this acceleration is independent of the body's mass and is purely a result of the coordinate transformation between the inertial and the rotating non-inertial frames.

Let us work out the details of the centrifugal and Coriolis accelerations from scratch in the context of Newton's world. Later we will revisit rotating reference frames in the context of general relativity.

If a mass m is at rest on a turntable that is rotating in an otherwise inertial frame, the mass experiences a centripetal force; and if it is moving with respect to the turntable, it experiences a Coriolis force as well. To begin, pin the mass a distance r from the axis of the turntable. If the angular velocity of the turntable is ω, then the transverse velocity v_t of the mass is ωr, as shown in Figure 7.2.

FIGURE 7.2 ▶

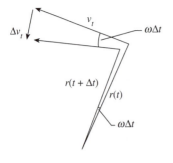

FIGURE 7.3 ▶

Because the direction of v_t is changing, the mass is subject to an inward acceleration, called the centripetal acceleration. We can calculate it by viewing Figure 7.3, which shows the velocity at time t and at time $t + \Delta t$. The angle between the two velocities in the figure is $\omega \Delta t$, so the change in the velocity is $v_t \omega \Delta t$ and its direction is inward. So,

$$dv_t/dt = v_t \omega = r\omega^2 = v_t^2/r,$$

which is the familiar expression for the centripetal acceleration. If we release the mass abruptly, it accelerates outward, as viewed in the rotating frame, with an acceleration equal and opposite to the centripetal acceleration. This acceleration is called the centrifugal acceleration. When you swing a mass on a string, you feel the centrifugal acceleration pulling the string taut.

Next, recall the Coriolis acceleration. This acceleration acts perpendicular to the direction of motion of the mass relative to the turntable (to the right of the particle's velocity if ω is positive) and has a magnitude of $2v'\omega$, where v' is the velocity of m relative to the turntable. Let us derive this result from scratch. First we consider transverse motion, motion at a constant r around a circle, and then we look at radial motion.

So, consider a particle with a given transverse velocity v. Decompose v into two pieces: the velocity of the turntable at r, $v_t = \omega r$; and the velocity difference, $v' = v - v_t$, or the velocity of the mass m relative to the turntable at that point. The full centripetal acceleration is then

$$\frac{v^2}{r} = \frac{(v_t + v')^2}{r} = \frac{v_t^2}{r} + 2v'\omega + \frac{v'^2}{r}.$$

The second term on the right-hand side is the Coriolis acceleration. In order for the particle to remain on a circle of fixed r, its contact to the turntable must provide this additional inward acceleration, $2v'\omega$. The particle itself pushes on the turntable with a force $-2v'\omega m$, which is called the Coriolis force. The third term in the equation, v'^2/r, is the additional centripetal acceleration produced by the increased speed.

Next, let m have a radial velocity v' along an axis of the turntable. As the particle travels outward, its transverse velocity increases and its direction of motion changes, so there are two contributions to its acceleration. Because the angular velocity ω of the turntable is fixed, its transverse velocity is $v_t = \omega r$. But r changes because the mass has a radial velocity v' with respect to the turntable, $\Delta r = v' \Delta t$. So, the magnitude of the transverse velocity changes $\Delta v_t = \omega \Delta r = v' \omega \Delta t$, and we see that the first contribution to the acceleration, which is clearly perpendicular to its direction of motion, is $v' \omega$. The second contribution to the acceleration comes from the change in direction of v' as the mass m moves radially on the rotating turntable. Because the turntable rotates through the angle $\omega \Delta t$ in the time interval Δt, the mass develops an additional transverse velocity, $v' \omega \Delta t$. We see that both contributions to the particle's transverse velocity act in the same direction and sum to $2v' \omega$, as shown in Figure 7.4.

If we speak about forces, we see that the turntable pushes on the mass with a transverse force of $2v' \omega m$, so that it continues in a radial direction. So, if the particle is not attached to the turntable, it accelerates to the right of its direction of motion, as measured in the turntable frame, at a rate

$$a_{\text{Coriolis}} = -2v' \omega,$$

which is the well-known Coriolis acceleration result we sought.

In summary, for both radial and transverse velocities relative to the turntable, we have the same expression for the Coriolis acceleration. Because any velocity can be decomposed into a radial component and a transverse component, our result for the Coriolis acceleration is perfectly general.

What have we learned so far? We see that force-free straight-line motion in an inertial frame is interpreted as accelerated motion in a non-inertial frame. If we introduce forces to describe the accelerated motion in the non-inertial frame, we should use the term apparent for this reason. These

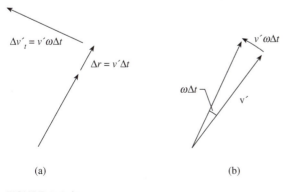

(a)   (b)

FIGURE 7.4 ▶

accelerations are determined purely by the relation of the inertial and the non-inertial coordinate systems, and can be derived without any knowledge of mechanics. All we need is geometry!

Now we come to the good stuff. Einstein argued that gravity is also an apparent force. Consider first a body moving in a constant gravitational field where it experiences a constant acceleration $-g$ independent of its mass. This problem is indistinguishable from the motion of a body in a frame free of gravity but accelerating upward at the rate g. These facts were well appreciated by classical physicists in Newton's era, but this statement of the Equivalence Principle was popularized and pursued in the context of the relativistic theory of space-time by Einstein. Einstein posed the equivalence—the impossibility of distinguishing the physics in a constant gravitational field from that in an accelerating frame—by imagining a physicist doing experiments in an elevator. The elevator is accelerating upward at a rate g, and, Einstein claimed, if the elevator has no windows so the physicist cannot see the tricks being played on him, there is no experiment he can run that can distinguish this environment from one at rest on the surface of a planet where gravity generates the acceleration g. (On the surface of Earth, Newton's gravitational force law gives $g = GM_e/R_e^2 \approx 9.8$ m/s^2, where R_e is the radius of Earth and we have neglected all other planetary masses.)

We come to the conclusion that gravity is an apparent force much like centripetal and Coriolis forces, familiar from our experiences with turntables. As with any apparent force, we can transform to another coordinate system where the apparent force vanishes identically. In the case of a constant gravitational field, we can just consider an observer in free fall in that environment. According to the Equivalence Principle, this free-falling frame of reference is force-free and is an inertial frame where isolated masses travel along straight lines.

The idea that gravity can be transformed away by passing to a free-falling, *inertial* frame of reference is very illuminating. Because we know that masses move along straight lines in inertial frames, we can use the Equivalence Principle to solve any mechanics problem in a given gravitational field—just consider the motion from the perspective of a freely falling frame where special relativity holds, solve the problem, and finally map it back to the coordinates an observer would use at rest in a gravitational field. (It is important here to check that there are no external nongravitational forces, such as electrostatics, in the environment. These forces are not apparent and cannot be transformed away by a slick choice of reference frame.)

But, as emphasized by Einstein, we can also go beyond ordinary mechanics problems because the Equivalence Principle applies to *any* process. An interesting application concerns the deflection of light by a gravitational field. Because light moves along straight lines in inertial frames,

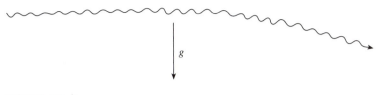

FIGURE 7.5 ▶

the Equivalence Principle implies that it experiences an acceleration when it moves transverse to a gravitational field, as shown in Figure 7.5. The deflection of light by the gravitational field of the Sun was computed using Newtonian mechanics in the early days of the nineteenth century. Einstein repeated the calculation, which we study in detail in a later section, in the context of relativistic space-time in 1919 and found that the Newtonian prediction is too small by a factor of two.

The experimental observation of the deflection of light in a gravitational field comes from a lensing effect, as illustrated in Figure 7.6. When light passes by a large astronomical object, it is attracted, as implied by the Equivalence Principle, and it is deflected as shown in the figure. The effect is observed by carefully measuring the background stars around the halo of the Sun during an eclipse and comparing those measurements to the position of the background stars when the Sun is in a different part of the sky. These measurements are difficult; only after data were accumulated over many years was it possible to rule in favor of the relativistic prediction. In fact, modern radio astronomy techniques using interferometry were the first to give decisive results with small enough errors to please the skeptics.

Finally, let us reconsider the scattering of light in the presence of a large star somewhat more critically. Because the direction of the gravitational acceleration varies as we move around the star, the use of the Equivalence Principle must be stated more carefully. As we pass around the Sun, the Equivalence Principle can be applied only *locally*; that is, only over a region of space-time where the gravitational field is essentially uniform can we find a freely falling inertial frame in which the field is essentially eliminated. The theory speaks of "local inertial frames" to accommodate spatially

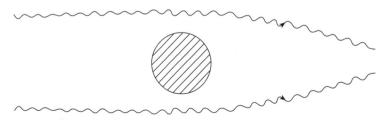

FIGURE 7.6 ▶

varying gravitational fields. We certainly cannot eliminate the effects of gravity from large space-time regions. The spatial dependence of gravitational fields means that the mathematical details of the theory change from point to point. For example, a body falling in a spatially varying gravitational field executes straight-line motion in each local inertial frame approximating the varying gravitational field. The actual trajectory of the body is obtained by patching together its simple trajectories in contiguous inertial frames. This sounds awfully complicated. Mathematically, the language for this motion is neatly given by differential geometry. The motion of the body is as simple as it can be in each inertial frame—just a straight line—but the relationship of the contiguous inertial frames must be described. The natural mathematical language is to describe space-time as curved and describe it with differential geometry. The motion of the body is as simple as possible— it travels on geodesic world lines, the shortest distance between two events in space-time. It will take several sections and developments to make these ideas precise and to obtain some classic predictions of general relativity, the gravitational red-shift, the resolution of the Twin Paradox, and the deflection of light in a gravitational field, which make up the bulk of this introductory course. Most of this work uses just our understanding of special relativity. We will not delve into the field equations of General Relativity, which requires more background than we have and more mathematical experience, but the Equivalence Principle serves us well and provides a solid introduction to the subject. Gravitational waves and all that are, alas, beyond the scope of this book.

In fact, the only part of general relativity we can cover using the Equivalence Principle and our knowledge of special relativity consists of situations where the gravitational field is not too strong. In most of our applications, we consider bodies with $v/c \ll 1$, so Newtonian kinematics suffice to leading order. Although this limits the range of topics we can approach, much of it is accomplished in keeping with the spirit of this book—the algebra and mathematical complexity are kept at a minimum, while the physics is, hopefully, particularly clear and the arguments complete and digestible.

Clearly, most of general relativity will remain in the student's future even after mastering this introduction. As R. P. Feynman was fond of saying [8], "It is good to have things to look forward to."

7.2 | Motion in a Rotating, Relativistic Reference Frame

In special relativity we plot the world line of a particle's motion on a Minkowski diagram (Figure 7.7). In general relativity we use notions such as the invariant interval, proper time, and proper length more than we do

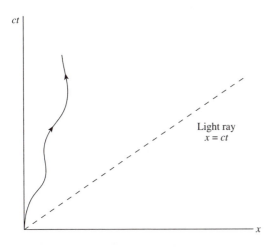

FIGURE 7.7 ▶

in our discussions of special relativity. So, let us review these ideas briefly and see that we can read effects such as time dilation, Lorentz contraction, and the simultaneity of relativity from expressions for the invariant interval. A problem familiar from elementary mechanics, such as the rotating reference frame, helps us move gradually into new, uncharted subjects.

Recall that the proper time $d\tau$ that passes on a clock attached to the moving particle can be calculated from the invariant interval ds,

$$ds^2 = c^2 dt^2 - dx^2 - dy^2 - dz^2. \tag{7.2.1}$$

Let us review the meaning of the symbols in this formula. In a frame S, we imagine two events, one at (ct, x, y, z) and another at $(ct + c\,dt, x + dx, y + dy, z + dz)$. ds is then the invariant interval between them. The events might be ticks on a clock, measurements of the ends of a rod, or whatever. The crucial point is that ds is the *same* in all reference frames, as we discussed in Section 4.3, so if we can calculate and understand it in one frame, we have its value in *all* frames. So, the proper time $d\tau$ that passes on a clock attached to the moving particle can be computed from ds by boosting to a frame S′ where the particle is at rest. In S′, $dt' = d\tau$ (proper time), $dx' = dy' = dz' = 0$, so $ds^2 = c^2 d\tau^2$. This means that if the particle is moving along the x axis at velocity v, so $dx = v \cdot dt$, $dy = 0$, and $dz = 0$, then Eq. (7.2.1) reduces to

$$ds^2 = c^2 d\tau^2 = c^2 dt^2 - v^2 dt^2 = (c^2 - v^2) dt^2$$

or

$$d\tau = \sqrt{1 - v^2/c^2}\, dt \equiv dt/\gamma,$$

which is just the expression of time dilation—the moving clock runs slowly.

Minkowski diagrams, space-time pictures, are the natural arena for discussing dynamics because they show time and position information together. Because space and time mix under boosts, we must work in four-dimensional space-time. If a particle's world line passes through $P_1 = (ct_1, x_1)$ and $P_2 = (ct_2, x_1)$, we know its velocity $v = (x_2 - x_1)/(t_2 - t_1)$ and its path.

As a first step toward developing relativistic particle motion in a gravitational field, let us consider relativistic force-free motion in a rotating reference frame. Because the forces in rotating frames, centripetal and Coriolis, are apparent just like gravity, this exercise in familiar territory is worth doing well. Choose a cylindrical spatial coordinate system,

$$x = r \cos \varphi, \quad y = r \sin \varphi, \quad z = z$$

as shown in Figure 7.8. Then the spatial distance becomes

$$dx^2 + dy^2 + dz^2 = dr^2 + r^2 d\varphi^2 + dz^2$$

and the space-time invariant interval reads

$$ds^2 = c^2 dt^2 - (dr^2 + r^2 d\varphi^2 + dz^2).$$

To describe in this language a turntable that is rotating about the z axis at angular velocity ω, as in Figure 7.9, we introduce a new azimuthal angle φ',

$$\varphi' = \varphi - \omega t, \tag{7.2.3}$$

so a point with fixed φ' has its φ increasing as ωt. This simple equation mixes the time coordinate with a spatial coordinate, so

$$d\varphi' = d\varphi - \omega dt$$

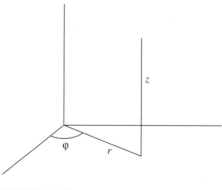

FIGURE 7.8 ▶

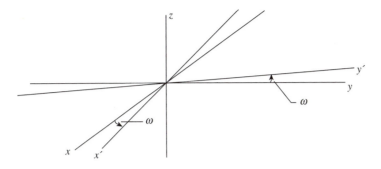

FIGURE 7.9 ▶

and the invariant interval written in terms of the rotating coordinates becomes

$$ds^2 = (c^2 - \omega^2 r^2)dt^2 - (dr^2 + r^2 d\varphi'^2 + 2\omega r^2 d\varphi' dt + dz^2) \quad (7.2.4)$$

and does not neatly separate into a spatial and temporal part.

Before dealing with ds^2 in its full glory, let us consider a clock at rest in the rotating reference frame and a distance r from the z axis. Two ticks of the clock occur at a given r, φ', and z, so $dr = d\varphi' = dz = 0$, and Eq. (7.2.4) reduces to

$$ds^2 = c^2 d\tau^2 = (c^2 - \omega^2 r^2)dt^2.$$

So, the proper time kept by the clock is

$$d\tau = \sqrt{1 - \omega^2 r^2/c^2}\, dt. \quad (7.2.5)$$

This result is the usual time dilation formula because the velocity of the clock relative to the inertial x, y, z frame is $v = \omega r$—as r increases, the clock's velocity (transverse) increases proportionally and the time dilation effect is enhanced.

Now let us deal with the new feature of Eq. (7.2.4)—the cross term, $-2\omega r^2 d\varphi' dt$, which mixes time and space intervals. To appreciate what is going on, we return to basics and imagine setting up a cylindrically symmetrical gridwork of rods and synchronized clocks to make position and time measurements in the rotating reference frame. First, consider two clocks having the same r and φ' but different z values, as shown in Figure 7.10. The clocks are rotating about the z axis at an angular velocity ω. We place a signal generator halfway between them, at rest in the rotating frame, to synchronize them. Because clocks 1 and 2 have velocities transverse to z, they receive the light pulses simultaneously in both the fixed (x, y, z) inertial frame and the rotating frame. The synchronization procedure produces clocks that are synchronized in both frames. Similarly, clocks that have the

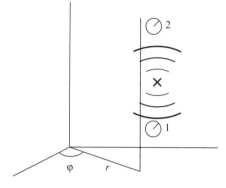

FIGURE 7.10 ▶

same φ' and z but slightly different r values are synchronized in both frames by our usual procedure.

However, clocks at the same r and z but different φ' suffer a different fate, as illustrated in Figure 7.11, where the view is from above. We have placed our signal generator halfway between the two clocks at their common r and z values. But wait! The signal generator and the two clocks are each moving in separate directions and each is experiencing a different centrifugal acceleration. We do not know how to handle this sort of situation using the principles of special relativity and the Equivalence Principle. However, if we move clocks 1 and 2 nearby, so they are separated by just an infinitesimal distance, we have a more secure, familiar situation. In this case, all three objects have essentially the same velocity and they can be viewed from a locally inertial frame moving at their common velocity. Because clocks 1 and 2 are separated in their direction of motion, we know from our discussion of the relativity of simultaneity

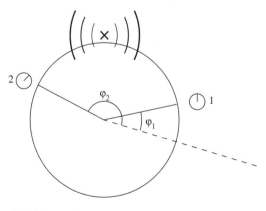

FIGURE 7.11 ▶

that they are *not* synchronized in the inertial frame defined by the coordinates (ct, x, y, z). The reason for this is clear—since light travels at the speed limit c with respect to any inertial observer, an observer at rest in the (ct, x, y, z) frame notes that clock 2 receives the light signal from the signal generator after clock 1 does because clock 2 is racing away from the source and clock 1 is racing toward the source. So, events that are synchronous in the rotating frame (clocks 1 and 2, for example) are not synchronous in the fixed inertial frame (ct, x, y, z) if they occur at different φ values. In fact, we know how large this effect is from our discussion of the relativity of simultaneity—it is the product of the velocity of the moving clocks times the distance between them in their rest frame divided by c^2. The relevant velocity is ωr. The relevant distance in the inertial frame (ct, x, y, z) is $rd\varphi$, which corresponds to a larger distance $\gamma(r) r d\varphi'$ in the clock's rest frame. Here $\gamma(r) = (1 - \omega^2 r^2/c^2)^{-1/2}$, so we have explicitly written γ as a function of r. (The notion of a local inertial frame is certainly important here.) So, $(\omega r)[\gamma(r) r d\varphi']/c^2$ is the time discrepancy that the frame (ct, x, y, z) notes on the clocks. But this is a time interval in the rotating frame, and we want the time difference in the fixed frame. This is given by multiplying by another factor of $\gamma(r)$ to account for time dilation, so the time difference is

$$\frac{\gamma^2(r) \omega r^2 d\varphi'}{c^2} = \frac{\omega r^2 d\varphi'}{c^2 - \omega^2 r^2}.$$

But the two clocks are synchronized in the rotating frame, so the time that passes there, dt', must be related to dt, the time that passes in the inertial frame, by

$$dt' = dt - \frac{\omega r^2}{c^2 - \omega^2 r^2} d\varphi'.$$

It must be that if we use this t' axis, the invariant interval is now split into a spatial part and a temporal part. Substituting $dt = dt' + \omega r^2 d\varphi'/(c^2 - \omega^2 r^2)$ into Eq. (7.2.4), we find

$$ds^2 = (c^2 - \omega^2 r^2) dt'^2 - \left(dr^2 + \frac{c^2 r^2 d\varphi'^2}{c^2 - \omega^2 r^2} + dz^2 \right) \qquad (7.2.6)$$

and everything has worked out fine.

In order to interpret Eq. (7.2.6), we can compare it to the invariant interval of an inertial frame of reference chosen to approximate the transverse velocity ωr locally. For example, taking a clock at rest in the rotating frame, $ds^2 = (c^2 - \omega^2 r^2) dt'^2$, and comparing it to a clock at rest in a locally inertial frame, $ds^2 = c^2 d\tau^2$, we have time dilation again, $d\tau = \sqrt{1 - \omega^2 r^2/c^2}\, dt'$. Similarly, taking a meter stick pointing in the transverse direction at radius r, $ds^2 = -c^2 r^2 d\varphi'^2/(c^2 - \omega^2 r^2)$ and comparing that to a meter stick pointing in the same direction in a locally inertial frame, $ds^2 = -r^2 d\varphi^2$, we have

$d\varphi = d\varphi'/\sqrt{1 - \omega^2 r^2/c^2}$, which is Lorentz contraction again because $r d\varphi$ is the proper length of the stick. These observations explain why the dt'^2 term in Eq. (7.2.6) has a prefactor $(c^2 - \omega^2 r^2)$ and the $d\varphi'^2$ term has that factor in the denominator. We see similar systematics in other invariant intervals in other applications for the same physical reason—the time dilation factor in special relativity is the same as the Lorentz contraction factor.

In order to have a broader perspective on the invariant interval let us turn to a short discussion of curved surfaces and non-Euclidean geometry.

7.3   A Look at Curved Surfaces

The language of general relativity is differential geometry, in both its classical and modern forms. Let us take a very informal look at some of these ideas because they play an important role in general relativity and modern theoretical physics.

Everyone knows how to set up Cartesian coordinates in a flat three-dimensional Euclidean space, as shown in Figure 7.12.

Now consider two points, P_1 and P_2, which are very close: one at (x, y, z) and the second at $(x + dx, y + dy, z + dz)$. The vector between the points is written as $d\mathbf{r} = \mathbf{i}\, dx + \mathbf{j}\, dy + \mathbf{k}\, dz$. The length squared of this infinitesimal vector is given by the dot product of the vector with itself, and using the fact that the unit vectors $\mathbf{i}$, $\mathbf{j}$, and $\mathbf{k}$ are mutually orthogonal we have

$$d\mathbf{r}^2 = dx^2 + dy^2 + dz^2,$$

which reproduces the Pythagorean theorem. This sort of space, called a metric space because it has a notion of distance, is Euclidean because the Pythagorean theorem holds. It is also called flat.

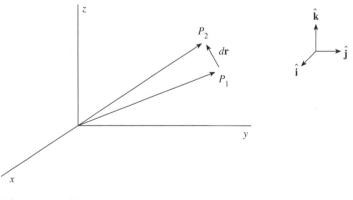

FIGURE 7.12 ▶

We can write the metric $d\mathbf{r}^2$ in any convenient coordinate system. For example, in the cylindrical coordinates discussed and illustrated in Section 7.2, the metric reads

$$d\mathbf{r}^2 = dr^2 + r^2 d\varphi^2 + dz^2.$$

This expression is more complicated than the expression for the metric in Cartesian coordinates, but the underlying geometry is the same old three-dimensional, flat Euclidean space. The point here is that when you work in differential geometry you must sometimes dig deeper than just the appearances of the formulas to uncover real intrinsic aspects of the space or surface you are studying. Do not be fooled by how expressions are parametrized. The emphasis in modern differential geometry is on intrinsic aspects of spaces and the relations between them. Modern notation, which we do not use here, avoids coordinates much more than classical approaches and just manipulates mathematical objects with real intrinsic geometric significance.

One such quantity is the length $\mathbf{r}^2$ itself. In the context of relativity, the invariant interval (or metric as we call it here from time to time) is such a quantity in our space-time world of Minkowski diagrams. For special relativity, the metric reads

$$ds^2 = c^2 dt^2 - dx^2 - dy^2 - dz^2 \qquad (7.2.1)$$

and it is important because of its invariance properties—it is the same in all inertial frames. These invariance properties follow from the fact that the speed limit c is the same in all frames. As we have seen, we can use the metric to derive time dilation, Lorentz contraction, and the Lorentz transformation laws. So, the invariance of ds^2 characterizes relativity in a very crisp fashion. Because most of us are particularly familiar with the Euclidean metric, which is always positive, it takes some care to call ds^2 a metric as well. For example, two space-time events that are connected by a light ray have $ds^2 = 0$, as we have discussed and illustrated before in Minkowski diagrams. If events are simultaneous but spatially separated in one frame, then their interval is negative and can be interpreted as a proper length. If, however, the two events are separated in time but occur at the same spatial point, their interval is positive and can be interpreted as a proper time.

It is instructive to consider curved surfaces embedded in three-dimensional Euclidean space. Consider the surface shown in Figure 7.13, the two-dimensional coordinate web on the surface (μ_1, μ_2), and the tangent vectors $(\mathbf{e}_1, \mathbf{e}_2)$. The vector between two nearby points on the surface can be written $d\mathbf{r} = dx^1 \, \mathbf{e}_1 + dx^2 \, \mathbf{e}_2$, where dx^1 and dx^2 represent the lengths along the coordinate directions $(\mathbf{e}_1, \mathbf{e}_2)$ on the surface (the superscripts on dx^1 and dx^2 denote directions, not algebraic powers). The distance between the points is

$$\mathbf{dr}^2 = \sum_{ij} \mathbf{e}_i \cdot \mathbf{e}_j \, dx^i dx^j,$$

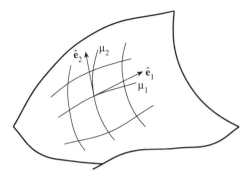

FIGURE 7.13 ▶

where the sum over the indices runs over $i = 1, 2$ and $j = 1, 2$. It is conventional to call the collection of dot products of the unit vectors, $\mathbf{e}_1$ and $\mathbf{e}_2$, the metric tensor, $g_{ij} = \mathbf{e}_i \cdot \mathbf{e}_j$. Note that because $\mathbf{dr}$ is confined to a curved surface where the unit vectors $\mathbf{e}_1$ and $\mathbf{e}_2$ are not orthogonal, the expression for $\mathbf{dr}^2$ has a cross term $2\mathbf{e}_1 \cdot \mathbf{e}_2 \, dx^1 \, dx^2$. However, there is little real geometric significance to all this algebraic complexity. For example, we can set up a coordinate patch on the surface where $\mathbf{e}_1$ and $\mathbf{e}_2$ are mutually orthogonal in the vicinity of a specific point—we say that we can choose a locally Euclidean coordinate system to describe the surface. But be careful. This coordinate system may be orthogonal in the vicinity of one point, but not retain that property as we move away. The generalization of these ideas to space-time in a gravitational field is important, as we show later.

It is clear that there is enormous freedom in choosing coordinate patches on a smooth surface. But how do we find true intrinsic properties of the surface without getting lost in all these matters of convenience and convention? This was the question posed by Carl Fredrick Gauss, the great mathematician who was one of the pioneers of the field. He asked the question: Is it possible to tell if a space or surface is Euclidean or is curved just from measurements within it, intrinsic measurements? Gauss answered the question yes and gave a definition of curvature that produces the same result no matter what local coordinates we use on the surface.

Take a familiar example—a sphere. This surface has constant positive curvature. What do we mean by that? Instead of giving a formal answer, let us contrast the features of the sphere to a flat two-dimensional space such as a plane (e.g., a piece of paper). If we have two points on a plane, the curve of shortest distance between them is a straight line. On the sphere, a curve of shortest distance is a great circle, called a geodesic, familiar from air travel around Earth. Using these geodesics, we can then form small, local, closed paths and compare their properties to familiar constructions on the plane. For example, the sum of the interior angles of a triangle drawn on a plane is 180°. If we make the same construction on the sphere using its geodesics or great circles, we clearly find that the sum of the interior angles is less than 180°. The deficit angle can be used as a definition of intrinsic curvature. Gauss liked this idea so much that he applied it to the space in which we live. He assumed that light travels on geodesics in our world (good!), and he used three mountains near him as vertices of his geodesic triangle. He found that the sum of the interior angles was 180° to good accuracy, so he concluded correctly that space in the vicinity of Earth is well approximated by Euclidean space.

An equivalent way of defining curvature is to consider a circle on the surface of interest and make it small, with a geodesic radius of distance a. Now measure the circumference, C. If a and C are related by a factor of 2π, then the surface is flat. If the circumference is less than that, the surface has positive curvature. And if the circumference is greater than that, the surface has negative curvature. Let us do this exercise using the sphere shown in Figure 7.14.

The circumference of the circle shown is $C = 2\pi R \sin(a/R)$, where R is the radius of the sphere. Now define the curvature to be proportional to the difference between $2\pi a$ and C,

$$K = \frac{3}{\pi} \lim_{a \to 0} \left(\frac{2\pi a - C}{a^3} \right).$$

Using the formula for the circumference and letting a be small so we can expand the sine, $\sin x = x - x^3/6 \ldots$ when $x \ll 1$, we find that $K = 1/R^2$, which justifies our definition. The result $K = 1/R^2$ shows that small spheres have more curvature than large ones (very reasonable). Clearly in the case of the sphere, the curvature doesn't change as we move around the surface, but other surfaces can have nontrivial functions $K(x^1, x^2)$. The student should check that a surface in the shape of a saddle has a negative curvature and

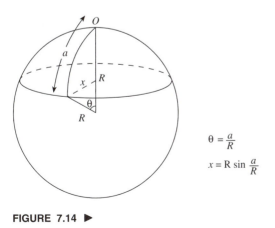

$\theta = \dfrac{a}{R}$

$x = R \sin \dfrac{a}{R}$

FIGURE 7.14 ▶

that a cylinder, which can be cut in the direction along its axis and laid flat on a plane, has a vanishing curvature.

Gauss and others have provided proofs for us that K is an intrinsic property of the surface and can be calculated from the components of the metric tensor g_{ij}, discussed previously. So g_{ij} is the source of fundamental intrinsic geometric features of the surface. After Gauss, other great mathematicians, such as Riemann, generalized these concepts of surfaces and geometry to metric spaces of any dimension and they dealt with intrinsic properties directly. For example, they discussed surfaces without embedding them in a Euclidean space of higher dimension—they dealt with curved space on its own terms. These developments were important precursors to general relativity. There is much, much more to this subject that requires more mathematics than we can handle, but we show the importance of the invariant interval (metric), locally flat tangent spaces, curvature, and geodesics in the material that follows. All these geometric concepts come up in the context of the four-dimensional space-time of the Minkowski diagrams we have already studied in the context of special relativity.

7.4 | Gravitational Red Shift

7.4.1 A Freely Falling Inertial Frame

The gravitational red-shift represents one of the simplest and decisive tests of the Equivalence Principle, which lies at the foundation of general relativity. We are interested in how light propagates in a gravitational field. So, suppose that light of frequency ν_e is emitted upward from the surface of a planet to be observed some distance h away. What frequency ν_o does the observer measure? We know that light travels at the speed limit c in any inertial frame and we know that a freely falling frame is inertial. So, consider first the propagation of the light signal from the surface of the planet from the perspective of an inertial frame falling freely in the uniform gravitational field g as shown in Figure 7.15. The light pulse travels a distance h in the time interval from $t = 0$ to $t = h/c$, and the falling frame has a velocity $v = gh/c$ downward relative to the observers at rest around the planet when the light pulse reaches its destination. The falling frame is perfectly inertial and the light pulse has frequency ν_e when it is emitted and when it is absorbed, as measured in the falling frame. But at $t = h/c$, the inertial frame is moving downward, opposite to the direction of propagation of the light pulse, so an observer at rest on the planet detects a Doppler shift toward the red when he observes the light pulse at $t = h/c$. The frequency shift is

$$\nu_o = \sqrt{\frac{1 - v/c}{1 + v/c}}\, \nu_e \approx \sqrt{1 - 2v/c}\, \nu_e \approx (1 - gh/c^2)\nu_e,$$

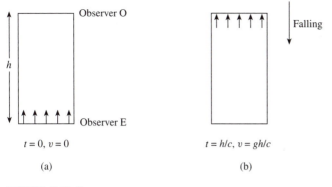

FIGURE 7.15 ▶

where we used Eq. (3.3.7) and $v = gh/c$, and linearized the formula for our present application, where $v/c \ll 1$.

The fractional change of the frequency is

$$\frac{\nu_o - \nu_e}{\nu_e} = \frac{\Delta \nu}{\nu} = -\frac{hg}{c^2}. \qquad (7.4.1a)$$

So, the observer a height h above the surface of the planet observes a slightly red-shifted wave.

Another way of presenting this result, which is particularly fundamental, is to say: *Identically constructed clocks run slower in lower gravitational potentials.* Because frequency varies as the reciprocal of time, $\nu = 1/t$, we can express Eq. (7.3.1a) as a fractional time difference,

$$\frac{\Delta t}{t} = \frac{hg}{c^2}. \qquad (7.4.1b)$$

In other words, an observer E at height 0 could send light signals to observer O at height h once a second. Eq. (7.4.1a) then states that observer O detects these signals more widely spaced in time, at a diminished or red-shifted frequency. So, observer O concludes that the clock at height 0 is running more slowly than hers, according to Eq. (7.4.1b).

Just as the observer at height 0 sent light signals to an observer at height h to compare clocks, their roles could be interchanged. Then the same argument, modified to account for the fact that now the freely falling frame is accelerating in the same direction as the light ray, so the Doppler shift formula applies with v replaced by $-v$, predicts that the lower observer detects a blue-shifted light ray. The lower observer concludes that the clock in the higher gravitational potential runs faster than his by a fractional change of $-hg/c^2$.

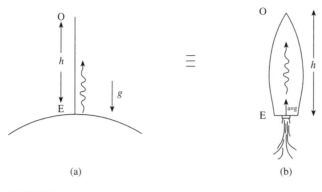

FIGURE 7.16 ▶

7.4.2 An Accelerating Spaceship

Another way to use the Equivalence Principle to analyze the gravitational red-shift is to replace the problem in a gravitational field with one in an accelerating reference frame (e.g., a spaceship), as shown in Figure 7.16. The spaceship is in a background inertial frame, so light travels at velocity c there regardless of its origins. So, light leaves the base of the ship (E), travels up during a time $t = h/c$, and is received at the tip of the ship (O). But O is receding at velocity $v = at = gh/c$, so it detects the frequency of light

$$\nu_o = \sqrt{\frac{1 - v/c}{1 + v/c}}\,\nu_e \approx \sqrt{1 - 2v/c}\,\nu_e \approx (1 - gh/c^2)\nu_e$$

in agreement with our earlier argument.

The light sent from E might be from a clock that sends pulses every second. So, O is observing how E's clocks work. Observer O detects $\Delta t/t = hg/c^2$ and she states that O's clocks run slowly because E is in a lower gravitational potential. So, we should understand the gravitational red-shift as a problem in clocks observed between frames. Consider a clock 2 at height h above a clock 1 at height 0. We know that O measures E's clocks as running slow because of the Doppler effect. Similarly, observer O can send light pulses every second and observer E can observe them. Using the equivalence of this problem to the accelerating spaceship, we see that E observes a frequency ν_e, which is related to the frequency sent by observer O,

$$\nu_e = \sqrt{\frac{1 + v/c}{1 - v/c}}\,\nu_o \approx \sqrt{1 + 2v/c}\,\nu_o \approx (1 + gh/c^2)\nu_o$$

in agreement with our previous discussion.

7.4.3 Gravitational Red Shift and the Relativity of Simultaneity

The gravitational red-shift is certainly more fundamental than the Doppler shift formulas, so let us undertake the challenge to derive the result directly from the most basic features of relativity. Observer O wants to compare the amount of time that passes at her height h in the gravitational field to the amount that passes at height 0 where observer E resides. Suppose that O has a clock; call it clock 2. Observer E also has a clock with him, call it clock 1, but suppose that light beams cannot be sent between the two observers. How can the observers compare the operation of clocks 1 and 2? Well, observer O could drop clock 2 to observer E, and the clocks could be compared at height 0. This is a good idea because, when the observer O drops her clock 2, clock 2 is freely falling in gravity and it is in an inertial frame of reference where space and time follow the rules of special relativity that we know so well. (In fact, suppose as usual that there is a synchronized gridwork of clocks in this inertial, freely falling frame. We will need to consult them later in this argument.) It is important that all the clocks involved be constructed identically, as usual. Observer E at height 0 can then note the hands on clock 2 when it reaches him at time t. The time t is determined by the fact that the falling clock accelerates at a rate g through a distance h, so $gt^2/2 = h$. So, $t = \sqrt{2h/g}$ and the velocity of clock 2 when it passes observer E at height 0 is $v = gt = \sqrt{2hg}$. But observer E is not really interested in the time on clock 2; he wants to know how much time has passed at height h. Observer O at height h can look at a freely falling clock at height h that was synchronized with clock 2 in that freely falling inertial frame, and read off the amount of time that has passed. Now for the crux of the matter! Observer E, being a good student of relativity, knows that a freely falling clock at height h is *not* synchronized with clock 2 when measured from *his* perspective because of the relative motion between the frame where he is at rest, the surface of the planet, and the freely falling frame—the relativity of simultaneity states that E measures such a clock to be ahead of clock 2 by a time interval xv/c^2, where x is just the height h. But $x = vt/2$ because the acceleration is constant, so the time difference is $xv/c^2 = v^2t/2c^2$. Therefore, observer E states that a time $t + v^2t/2c^2$ has passed at height h while a time interval t has passed at $h = 0$. The total time interval T detected by observer E is, then,

$$T = t + \frac{v^2t}{2c^2} = \left(1 + \frac{v^2}{2c^2}\right)t = \left(1 + \frac{1}{2} \cdot \frac{2hg}{c^2}\right)t = \left(1 + \frac{hg}{c^2}\right)t.$$

So, observer E concludes that a clock at height h runs more quickly—more time has passed there—with a fractional difference of $\Delta t/t = hg/c^2$. In other words, observer E concludes that clocks at higher gravitational potential run faster. This result is in agreement with our previous conclusion. The relativity of simultaneity, the fact that clocks that are synchronized in one inertial frame are measured to be out of synchronization when measured by an observer in relative motion, is the fundamental idea behind the gravitational red-shift, correctly concludes observer E.

7.4.4 A Rotating Reference Frame

Our exercise in rotating reference frames affords yet another example of the gravitational red-shift. The proper time interval $d\tau$ was related to the time interval that passes in the fixed inertial frame as $d\tau = \sqrt{1 - \omega^2 r^2/c^2} dt$. A mass at rest in the rotating frame experiences a centripetal acceleration $\omega^2 r$, which is generated from a potential $V(r) = -\omega^2 r^2/2$, because $\omega^2 r = -(d/dr)(-\omega^2 r^2/2)$. So, we can write $d\tau = \sqrt{1 + 2V(r)/c^2} dt$ and the proper time intervals at different radii r_1 and r_2 are related,

$$\frac{d\tau_1}{d\tau_2} = \sqrt{\frac{1 + 2V(r_1)/c^2}{1 + 2V(r_2)/c^2}}.$$

Therefore, the frequency ν_1 of an oscillator at r_1 is related to that at r_2 by

$$\frac{\nu_1}{\nu_2} = \sqrt{\frac{1 + 2V(r_2)/c^2}{1 + 2V(r_1)/c^2}}.$$

If both $V(r_1)/c^2$ and $V(r_2)/c^2$ are much less than unity, we can linearize this expression and find

$$\frac{\nu_1}{\nu_2} \approx \left[1 + \frac{V(r_2) - V(r_1)}{c^2} \right]$$

or

$$\frac{\nu_1 - \nu_2}{\nu_2} \approx \frac{V(r_2) - V(r_1)}{c^2}. \tag{7.4.4a}$$

In the case of the rotating reference frame, $V(r) = -\omega^2 r^2/2$. So, if a pulse of light is emitted radially from r_1 and is observed in the rotating frame at $r_2 \neq r_1$, the observer at r_2 detects a frequency-shifted pulse. Clearly, this is just an example of the transverse Doppler shift of special relativity because $v_t = \omega r$ is the transverse velocity of the rotating reference frame at radius r.

A general discussion of the gravitational red-shift in a non-uniform gravitational field V predicts

$$\frac{\Delta\nu}{\nu} \approx -\frac{\Delta V}{c^2}, \tag{7.4.4b}$$

where $\Delta\nu$ is the frequency difference and ΔV is the potential difference between the point of detection and emission of the light wave.

7.4.5 A Famous Experimental Test of Gravitational Red Shift

It is interesting that the best experimental determination of gravitational red-shift comes from terrestrial controlled experiments that rely on the Mössbauer effect to measure tiny frequency shifts rather than from observations of distant stars. Recall the Mössbauer effect discussed in Section 6.5. The experiment involved considering an emitting atom within a regular crystal array. The emitting atom shares its recoil energy and momentum with its entire crystal environment, so its recoil velocity is reduced to essentially zero and all the energy between the quantum energy levels in the atom shows up as the energy of the emitted light wave. If the light emitted from such a crystal is incident on another identical crystal, absorption is possible because the energy of the light exactly matches an energy difference in the spectrum of the receiving atom, which can absorb the light without recoiling. If either the absorbing crystal or emitting crystal has a small relative velocity, the attendant Doppler shift shifts the energy of the light enough to make resonant absorption impossible. (Recall that in quantum physics, the frequency of light is related to its energy by the Planck relation $E = h\nu$, where h is Planck's constant, $h = 6.627 \cdot 10^{-34}$ J-s, so we can speak equally well in terms of frequency or energy when discussing emission and absorption of photons by atoms.) In fact, the energy levels of atoms are not infinitely precise—each energy level has a natural width that is a consequence of the Heisenberg uncertainty relation. Mössbauer was then able to study the profiles of spectral lines, obtaining their widths and detailed shapes by using his crystals and the physics of the Doppler shift.

R. V. Pound, G. A. Rebka, and J. L. Snider used the Mössbauer effect to measure the red-shift of light near the surface of Earth [9]. They placed a source of ^{57}Co 22.6 m higher than a receiver. The Equivalence Principle predicts a fractional frequency shift of $\Delta\nu/\nu_e \approx 2.46 \cdot 10^{-15}$, which is very tiny: three orders of magnitude smaller than the intrinsic natural uncertainty in the frequency of the emitted light $\Delta\nu_{QM}$, $\Delta\nu_{QM}/\nu \approx 1.13 \cdot 10^{-12}$. In order to obtain an observable effect, they gave the source a sinusoidal motion $v = v_o \cos \omega t$ to induce a controlled, mechanical Doppler shift, $\Delta\nu_D/\nu = -(v_o/c)\cos \omega t$. Then the resonance-absorption cross section depends on $\Delta\nu + \Delta\nu_D$, and, if $\Delta\nu_D \gg \Delta\nu$, an exercise in quantum mechanics shows that the cross section has a term linear in $\cos \omega t$ with a strength proportional to $\Delta\nu$, the gravitational red-shift. By isolating this characteristic external frequency ω in the cross-sectional measurements, the experimenters confirmed the prediction of general relativity to good precision, about 10%. It is interesting that this terrestrial experiment is much more decisive than astronomical data of light emitted from distant stars because of all the inherent uncertainties in such observations (the Doppler effects of the movement of the star and the movement of the emitting atom in the turbulent hot surface of the star must be accounted for somehow).

7.4.6 Gravitational Red Shift and Energy Conservation

Let us end this discussion with a simple illustration of gravitational red-shift that focuses on the consistency of relativity with other basic principles of physics. We show that the gravitational red-shift is absolutely necessary for the internal consistency of the theory—without it we could construct a perpetual motion machine!

Consider a pulley on the surface of Earth supporting two observers at the ends of a string as shown in Figure 7.17 [10].

Let the difference of heights of the observers be h, as usual. Let the lower observer shine a flashlight at the higher observer, who sees the light, so it is absorbed on his retina. (Notice the similarity of this argument to our original derivation of $E = mc^2$, except the apparatus is now vertical and in an ordinary gravitational field.) Let the light have energy E, so from special relativity we know that it has a mass equivalent of E/c^2. So, if the light travels from the lower to the higher observer, then mass E/c^2 has been transferred between the observers; the upper one is now slightly heavier than the lower one, so it sinks and does work equal to the change in the potential energy, $hg \cdot E/c^2 = E \cdot \Delta V/c^2$. Our pulley system is now back to its original configuration, and we can repeat the process as many times as we wish and have an inexhaustible source of work.

We have done it! We have made a perpetual motion machine! Well, except for one thing—we forgot about the gravitational red-shift! The light detected by the higher observer in Figure 7.17 has a smaller frequency, given by our gravitational red-shift formula, $\Delta \nu/\nu = -\Delta V/c^2$. How can this get us out of our conundrum? If we return to Eq. (6.20), it is easy to see (as we verify later) that the energy that light carries transforms between frames exactly as its frequency. If that is the case, then the energy the light deposits on the retina of the higher observer is diminished by $E \cdot \Delta V/c^2$, which is

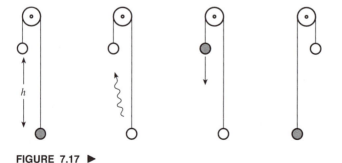

FIGURE 7.17 ▶

exactly the work we were hoping to get out of our machine! So, in reality, when the light from the lower observer reaches the higher one, its energy is diminished by the gravitational red-shift by just the amount we wanted to generate. So, yet another perpetual motion machine design bites the dust!

Let us check that light's frequency and energy transform identically in special relativity. Recall from Eq. (6.20) that if we know the energy E and x component of the relativistic momentum p_1 in frame S, then the energy in frame S' is $E' = \gamma(E - vp_1)$. But for light propagating in the x direction, $p_1 = E/c$, so the transformation law becomes

$$E' = \gamma\left(E - \frac{vE}{c}\right) = \gamma\left(1 - \frac{v}{c}\right)E = \sqrt{\frac{1 - v/c}{1 + v/c}}E,$$

which we recognize as the Doppler shift formula for the light's frequency. We learn that our scheme for making a perpetual motion machine fails miserably because relativity is perfectly consistent.

The fact that light's frequency ν and energy E transform identically under boosts is important in the quantum theory of light we have touched on in our discussion of relativistic collisions and quantum energy levels. Planck's equation, $E = h\nu$ (where h is Planck's constant), is consistent with special relativity because E and ν share the same transformation law.

This observation suggests yet another way of viewing the gravitational red-shift—it is just a consequence of energy conservation! When a photon, a quantum of light energy, travels from the observer on the surface of a planet where its frequency is ν_e and its height is 0, to an observer where its frequency is ν_o and its height is h, the total energy, accounting for the gravitational potential, must be conserved:

$$E_e = E_o + hg\frac{E_o}{c^2}.$$

The change in potential energy has been written as $hg(E_o/c^2)$, which is the change in the potential hg in the uniform gravitational field times the mass equivalent of the energy E_o there. Solving for E_o,

$$\frac{E_o}{E_e} = \frac{1}{1 + hg/c^2} \approx 1 - \frac{hg}{c^2},$$

which we can write as a fractional change in energy, which is also the fractional change in frequency,

$$\frac{E_o - E_e}{E_e} = \frac{\nu_o - \nu_e}{\nu_e} \approx -\frac{hg}{c^2},$$

and we have derived Eq. (7.4.1a) again from a different perspective!

This result gives us a nice alternative view of gravitational red-shift. Why does the frequency of light change as it propagates away from the surface of a celestial body? Because it propagates to a new location where its gravitational potential is larger, so its relativistic energy, and hence its frequency, must be diminished accordingly!

7.5   The Twins Again

When we last left our twins, Mary and Maria, they were coping with the fact that Mary is four years older than Maria after her space trip. In that discussion, we found that from Maria's perspective Mary ages an unexpectedly large amount when Maria jumps from the outgoing to the incoming rocket at the midpoint of her trip. Recall from that discussion that the lines of constant time in the frame of the outgoing rocket are at widely different angles from the lines of constant time in the frame of the incoming rocket, as shown in the Minkowski diagram, Figure 3.19. In fact, we saw that Maria measures that 6.4 years passes on Mary's clock during Maria's turnaround!

By using the device of two rockets, we have been able to analyze the twin paradox without explicitly considering space and time measurements in an accelerated reference frame. Now that we know that accelerated reference frames are equivalent to environments in a gravitational field, we can face the problem head on. Our only limitation is that our discussion of general relativity is good only for weak gravitational fields, or, equivalently, small values of v/c. To do better, we need the full apparatus of general relativity— Einstein's field equations. Nonetheless, we can make a small but meaningful step in the right direction at this level [10].

So, we choose to view the trip from Maria's frame. From Maria's perspective, Mary goes out and back. By the Equivalence Principle, then, Maria's acceleration can be replaced by a gravitational field. We already know that the important portion of the trip is the period during which Maria detects Mary's reversal. Call the distance between the sisters vT at this point, where T is the time they have been traveling apart at velocity v. (We must take $v/c \ll 1$ in this example, contrary to our previous discussion, because we only know how to calculate first-order general relativistic effects. If $v/c \ll 1$, then it does not matter whether Mary or Maria measures T—both observers find the same time to $O(v^2/c^2)$.) At this point, the turnaround, Maria actually turns on her rocket motors and experiences an acceleration a for a time t. However, invoking the Equivalence Principle, we can replace the rocket environment by one in which there is a gravitational field in Maria's frame that produces an acceleration a, as shown in Figure 7.18. The gravitational potential difference between the sisters is distance times acceleration, avT, which causes Mary's clock (her heartbeat) to gain the total amount $(avT/c^2)t$, according to our formulas developed for the gravitational red-shift in Section 7.4, Eq. (7.4.1b). Finally, t must be

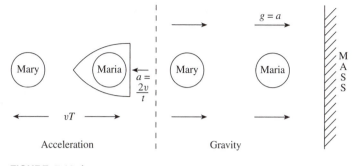

FIGURE 7.18 ▶

long enough to accommodate the reversal from the outgoing speed v to the incoming speed $-v$, so $at = 2v$. Therefore, the total time gained by Mary's clock is $2Tv^2/c^2$.

Does this result agree with our previous analysis of the Twin Paradox? Return to Figure 3.19 and compute the time that Maria states passes on Mary's clock during the turnaround. We pull out our trusty relativity of simultaneity formula, which reminds us that Maria measures clocks at rest in Mary's frame that are separated by a distance x to be out of synchronization by an amount xv/c^2. So, if we change from the outgoing to the incoming rocket, causing v to change by Δv, the time that passes on Mary's clock is $x\Delta v/c^2$. But x is the distance between the sisters in Mary's frame, which is vT in our discussion here. And finally, because the velocity v reverses, $\Delta v = 2v$. So, our earlier relativity of simultaneity discussion predicted that Mary's clock, as measured by Maria, gains a time of $2Tv^2/c^2$, in complete agreement with the general relativity result obtained here! Using the numbers of our previous discussion in Section 3.4, Figure 3.19, $T = 5$ and $v/c = 0.8$, so Mary's time gain is $2 \cdot 5 \cdot 0.8^2$, which is 6.4 years, as we have found earlier.

Note that our general relativity discussion was really only good for $v/c \ll 1$, so comparing it to our earlier discussion with $v/c = 0.8$ was somewhat arbitrary. What is important, however, is the fact that we obtained the same formula $2Tv^2/c^2$ with the same functional dependences from both arguments—nice!

7.6 | Making the Most out of Time

Let us take a brief interlude from our serious studies and illustrate some features of general relativity and the gravitational red-shift in an unusual setting. The gravitational red-shift means that clocks run quickly in high gravitational potentials and they run slowly in low gravitational potentials.

Suppose there is a colony of tough but gentle space creatures, called Scruffs, who live in a region of the universe where the gravitational potential varies rapidly in space. The Scruffs, who live in nuclear family units, are able to arrange their environment to their needs. In fact, they use the gravitational potential to control time for their day-to-day convenience. For example, each family of Scruffs has a room in its house where the gravitational potential is very negative. Scruffs call this room "time out," and whenever their baby boy is naughty, they give him a "time out" by putting him unceremoniously into this dreaded place. When the baby is put into "time out," all his actions slow down and are much more tolerable. Even the baby's whining is now heard at a lower, more tolerable frequency. This is a parent's dream come true!

The tricks do not end here. In the attic, each family has a "play room" where the gravitational potential is very high. Scruff teenage daughter, who is always behind on her homework, jumps into this "play room" when she needs extra time to finish her physics homework before its due date. Her father uses the room when he needs extra time to complete a project for his pushy boss.

Actually, life in a Scruff house can be quite stressful. Because there are regions where the gravitational potential varies rapidly in space, Scruffs experience strong forces as they move about. For example, as they put their noisy son into "time out," the gravitational potential varies rapidly across his body, pulling one side much more than the other. (These effects are called tidal forces and are similar to, but highly magnified compared to, the forces the Moon exerts on Earth. These tidal forces—the fact that the Moon pulls the side of Earth nearest to it more than the side further from it—flatten Earth slightly and cause high and low tides, a familiar effect that was first calculated by Isaac Newton.) Luckily, the Scruffs have evolved into tough little beings who can withstand these stretching and compressing forces. Even their pets take advantage of their environment. The family jat, a long and lazy creature, naps with its head in the "play room," so it can have a relatively long snooze, while its belly is in "time out" so it can sa'vor its lunch. The only problem with the jat is that no one can tell how old it is, because from the Scruffs' perspective, its various parts are aging at alarmingly different rates!

7.7   Gravitational Field of a Spherical Mass—The Schwarzschild Metric

The application of general relativity to ordinary planetary dynamics requires the gravitational field outside a spherical mass distribution representing the Sun. This is the relativistic version of Newton's gravitational potential,

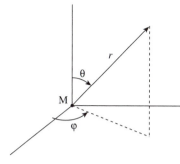

FIGURE 7.19 ▶

$-GMm/r$. In the language of general relativity, we need an expression for the invariant interval ds^2 outside a spherical mass distribution. We certainly want to use spherical coordinates for such a problem r, θ, and φ. Then ds^2 should have the form

$$ds^2 = H(r)c^2 dt^2 - J(r)dr^2 - (r^2 d\theta^2 + r^2 \sin^2 \theta d\varphi^2), \qquad (7.7.1)$$

referring to the coordinate system shown in Figure 7.19. We have incorporated unknown functions of r, H, and J in this expression to accommodate the effects of the spherically symmetrical gravitational field having a central symmetry point at the origin $r = 0$. The $d\theta$ and $d\varphi$ pieces of the invariant interval are written in the same way as for an inertial reference frame because they are lengths perpendicular to the radial gravitational field and should be unaffected by its presence. To determine the functions $H(r)$ and $G(r)$ rigorously, we really need to develop the field equation of general relativity, the relativistic versions of the Poisson equation of Newtonian gravity. We can not do that here, so we will be content with qualitative, intuitive arguments based on our experience with the gravitational red-shift, the rotating reference frame, and the Equivalence Principle. In this way, we can find H and J correct to first order in the Newtonian potential GM/r.

It is instructive to recall the invariant interval for the rotating reference frame,

$$ds^2 = \left(1 - \frac{\omega^2 r^2}{c^2}\right)c^2 dt'^2 - \frac{r^2 d\varphi'^2}{(1 - \omega^2 r^2/c^2)} - dr^2 - dz^2. \qquad (7.2.6)$$

Recall that the transverse velocity of a point at rest in the rotating frame and a distance r from the axis of rotation, the z axis, is $v_t = \omega r$. Therefore, the factor $(1 - \omega^2 r^2/c^2)$ in the $c^2 dt'^2$ term can be interpreted as generating the expected time dilation effect and the $(1 - \omega^2 r^2/c^2)^{-1}$ in the $r^2 d\varphi'^2$ term can be interpreted as generating the expected Lorentz contraction effect. In addition, because the centripetal force can be derived from the centripetal potential $V(r) = -\omega^2 r^2/2$, we can write Eq. (7.2.6) in the form

$$ds^2 = \left(1 + \frac{2V(r)}{c^2}\right)c^2 dt'^2 - \frac{r^2 d\varphi'^2}{(1 + 2V(r)/c^2)} - dr^2 - dz^2. \qquad (7.7.2)$$

Our discussion of the red shift in the presence of a uniform gravitational field can also be made in the context of the invariant interval. Now the gravitational forces are just in the vertical z direction, and we can incorporate our results of Section 7.4 in the expression

$$ds^2 = \left(1 + \frac{2gh}{c^2}\right)c^2 dt^2 - \frac{dz^2}{(1 + 2gh/c^2)} - dx^2 - dy^2. \qquad (7.7.3)$$

From this expression, we read that clocks that are at a lower gravitational potential run slower than those in a higher gravitational potential. In addition, meter sticks that are at a lower gravitational potential and aligned along the z axis are contracted compared to those at higher gravitational potential.

Now we can return to Eq. (7.7.1) for the invariant interval in the gravitational field of a mass M. Recalling that the gravitational potential is $-GM/r$, we are led to the educated guesses for the functions $H(r)$ and $J(r)$,

$$H(r) = J^{-1}(r) = \left(1 - \frac{2GM}{c^2 r}\right), \qquad (7.7.4)$$

where G is Newton's constant. This famous result is called the Schwarzschild metric. As written in Eq. (7.7.4) the answer is exact and was originally obtained by solving Einstein's field equations, the relativistic versions of the Poisson equation for the gravitational field around a point mass. Our argument is no replacement for the original! In fact, our derivation is at best good for weak fields where $GM/c^2 r \ll 1$ and we have no right to expect accuracy beyond terms linear in $GM/c^2 r$. The exact Schwarzschild metric reads,

$$ds^2 = \left(1 - \frac{2GM}{c^2 r}\right)c^2 dt^2 - \frac{dr^2}{(1 - 2GM/c^2 r)} - r^2(d\theta^2 + \sin^2 d\varphi^2). \qquad (7.7.5)$$

7.8 | Bending of Light in a Gravitational Field

As we discussed earlier, the Equivalence Principle implies that *all* physical phenomena experience the acceleration of gravity. This point applies to wave phenomena, as well as to particles with rest masses. Newton understood this point, and the original quantitative calculation of the bending of light in a nonrelativistic setting was published in 1801. The relativistic calculation

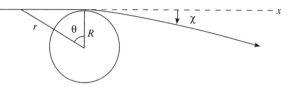

FIGURE 7.20 ▶

gives twice as large a deflection and is in agreement with modern high-tech experiments.

Before we review the relativistic calculation that uses the Schwarzschild metric, let us consider the nonrelativistic calculation. Consider a particle of rest mass m that glances by a star of mass M and radius R as shown in Figure 7.20. For a star similar to the Sun, the deflection angle χ proves to be very small. The particle feels the radial force of gravity and, as is clear from the symmetry in the picture, the star imparts a net transverse momentum to the particle, which bends its trajectory as shown. The calculation is successful only if m, the mass of the particle, cancels out of the calculation. The effect must be universal and independent of m. This occurs because, if we use Newton's theory of gravity, the particle experiences a force $\mathbf{f}$ proportional to its mass m which produces an acceleration given by $\mathbf{f} = m\mathbf{a}$, predicting an acceleration $\mathbf{a}$ independent of m. So, m is just scaffolding in the calculation. By considering accelerating reference frames instead of forces, we can calculate the deflection angle χ without ever needing the scaffolding m. Anyway, following the original Newtonian calculation done by Soldner (a German mathematician) in 1801, the gravitational force F produces a change in the particle's transverse momentum P_T,

$$dP_T = F \cos \theta \, dt. \tag{7.8.1}$$

We read from Figure 7.20 that $x = R \tan \theta$, so we can trade the linear position of the particle for the angle θ. The angle θ varies from $-\pi/2$ to $\pi/2$ during the process. In addition, because the deflection is small we have to good approximation $dx = cdt$, where we are also supposing that the particle's speed is very close to c. Because m cancels out of the calculation and we are interested in the $m \to 0$ limit, this is the right thing to do. Now, $dx = R\sec^2\theta d\theta$, so Eq. (7.8.1) becomes using $R = r \cos \theta$,

$$P_T = \int_{-\pi/2}^{\pi/2} \frac{GMm}{cR} \cos \theta \, d\theta = \frac{2GMm}{cR}. \tag{7.8.2}$$

The deflection angle is approximately

$$\chi = \frac{P_T}{P} = \frac{2GMm/cR}{mc}. \tag{7.8.3}$$

We see that m cancels here and the final answer is

$$\chi = \frac{2GM}{c^2R} \text{ radians.} \tag{7.8.4}$$

Substituting in the parameters for the Sun, we predict $\chi = 0.87$ seconds of arc, a tiny but measurable effect. Note that the effect is suppressed by two powers of the speed of light.

This curious calculation leaves several points unanswered. Treating light as a particle with rest mass and then taking the rest mass to zero at the end of the calculation is not really satisfying. In addition, light is a wave phenomenon so the applicability of this calculation is not really clear. The calculation works, actually, because of the universality of the effect—the details don't matter; what matters is that gravity produces a universal acceleration that affects all physical phenomena equally. All the different descriptions must get the same universal result.

To make this point, let us calculate the deflection of light by the gravitational attraction of the Sun by modeling light as a wave and treating space-time through general relativity [11]. Consider a light front propagating through a gravitational field described by the Schwarzschild metric. Our calculation will be accurate just to first order in $1/c^2$, appropriate for an ordinary gravitational field. Then, to good approximation,

$$ds^2 = \left(1 - \frac{2GM}{c^2r}\right)c^2dt^2 - \left(1 + \frac{2GM}{c^2r}\right)dr^2 - r^2(d\theta^2 + \sin^2\theta \ d\varphi^2). \tag{7.8.5}$$

Now let us specialize to the propagation of light. In an inertial reference frame, light travels on straight lines at the speed limit c. Along the world line, the invariant interval ds^2 vanishes. But because ds^2 is an invariant, $ds^2 = 0$ holds in the rest frame of the Sun as well as in a freely falling inertial frame. So, inspecting Eq. (7.8.5) with $ds^2 = 0$, we see that the velocity of light, as measured on the axes ct, x, y, and z, will be r dependent. So, when

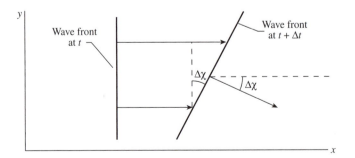

FIGURE 7.21 ▶

a wave front passes the Sun, different parts of it propagate at different speeds and the wave front changes direction. The situation is shown in Figure 7.21. The deflection $\Delta\chi$ is, reading from the figure,

$$\Delta\chi \approx \frac{v(y+\Delta y)\Delta t - v(y)\Delta t}{\Delta y} = \frac{\partial v}{\partial y}\Delta t.$$

So,

$$\frac{d\chi}{dx} \approx \frac{\partial v}{\partial y}\frac{dt}{dx} = \frac{1}{v}\frac{\partial v}{\partial y}.$$

The rate of change of $\partial v/\partial y$ is very small, of order $1/c^2$, so the deflection will also be very small. To calculate the deflection, we need the speed v to first order in $1/c^2$. To obtain this, write ds^2 in terms of dt and dx because the light rays are propagating in the x direction and deflection occurs only because $\partial v/\partial y$ is nonzero:

$$ds^2 = \left(1 - \frac{2GM}{c^2 r}\right)c^2 dt^2 - \left(1 + \frac{2GM}{c^2 r}\right)\left(\frac{dr}{dx}\right)^2 dx^2 - r^2\left(\frac{d\theta}{dx}\right)^2 dx^2.$$

Call the closest approach of the light ray to the Sun R as shown in Figure 7.22.

$$r = \sqrt{x^2 + y^2} \quad \cos\theta = \frac{y}{r}.$$

The derivatives we need in the metric, to effectively rewrite it in Cartesian coordinates, starting from polar coordinates, are

$$\frac{dr}{dx} = \frac{x}{r} \quad \frac{d\theta}{dx} = -\frac{y}{r^2}.$$

Substituting into the expression for the invariant interval,

$$ds^2 = \left(1 - \frac{2GM}{c^2 r}\right)c^2 dt^2 - \left(1 + \frac{2GM}{c^2 r}\right)\frac{x^2}{r^2}\,dx^2 - r^2\frac{y^2}{r^4}\,dx^2.$$

So,

$$ds^2 = \left(1 - \frac{2GM}{c^2 r}\right)c^2 dt^2 - \left(1 + \frac{2GMx^2}{c^2 r^3}\right)dx^2.$$

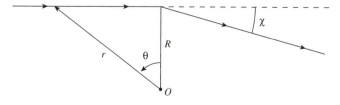

FIGURE 7.22 ▶

Setting $ds^2 = 0$ we obtain $v = dx/dt$, to first order in $1/c^2$,

$$v = \left[1 - \frac{GM}{c^2 r} \left(1 + \frac{x^2}{r^2} \right) \right] c.$$

To calculate the deflection, we need $(1/v)(\partial v/\partial y)$ to first order in $1/c^2$,

$$\frac{1}{v} \frac{\partial v}{\partial y} \approx \frac{GM}{c^2} \left(\frac{3x^2 y}{r^5} + \frac{y}{r^3} \right).$$

Integrating from $-\infty$ to $+\infty$ gives the full deflection,

$$\chi \approx \frac{GM}{c^2} \int_{-\infty}^{\infty} \left(\frac{3x^2 y}{r^5} + \frac{y}{r^3} \right) dx.$$

Integrate by using the variable θ, which varies from $-\pi/2$ to $\pi/2$. From Figure 7.22, this gives

$$x = R \tan \theta, \quad R = r \cos \theta, \quad y \approx R.$$

So,

$$\chi \approx \frac{GM}{c^2 R} \int (3 \sin^2 \theta + 1) \cos \theta \; d\theta$$

$$\chi = \frac{4GM}{c^2 R}$$

which is twice the Newtonian result, as promised!

This result played a very important role in the historical development and acceptance of general relativity. Better and better measurements of χ are actively being pursued.

7.9   Closing Comments

Let us end this introduction to general relativity with some assorted observations.

Although the bending of light in a gravitational field might be technically challenging to observe under ordinary conditions, just the existence of this effect suggests many things. As we discuss in the chapters on special relativity, mass and energy are actually unified into one concept, relativistic energy. Where you have mass, you have energy, and vice versa, as stated in $E = mc^2$. So, if mass generates gravitational fields, then so must energy in any of its forms. In particular, light waves carry energy and this energy must generate a gravitational field that acts on all matter, far or near. But this means that when two light beams pass near each other, each produces a gravitational field that bends the trajectory of the other! In other words, the universality of gravity implies that light attracts light and light beams scatter from one another. This effect is unbelievably small under ordinary conditions, but there are important matters of principle here. For example, the theory of light is intrinsically nonlinear if we include gravitational effects in our thinking. Light rays do not just pass through one another. The idea of linear superposition that we all learn in classical electricity and magnetism is just a superb approximation in the case of electromagnetic fields. It is not a matter of law. (There are other, much larger effects that cause light rays to interact and scatter. The quantum theory of electrons, positrons, and photons predicts such effects, called Delbrück scattering, which is just barely observable using modern techniques.)

There is something very special in gravitational attraction. Because all masses attract one another, there is no way to hide from gravity! Contrast this situation with electricity and magnetism. In that theory, there are positive and negative charges that can screen one another. Imagine a positively charged impurity in a piece of material. The impurity attracts the electrons in the material and the electric field emanating from the impurity is screened. The thoroughness of the screening depends on the details of the material in which the impurity finds itself, but it is typically quite dramatic and the screened impurity's electric field might have no long-range component, a field that would decay at large distances such as $1/r^2$. We notice the thoroughness of electromagnetic screening in day-to-day situations— bulk material tends to be electrically neutral. Such screening phenomena are impossible in gravity; all masses (and therefore energies) attract one another. How then can an extended universe exist? Why does it not just collapse under its own weight? These are simple questions awaiting future answers!

This short introduction to general relativity does not do justice to the subject by a long shot! With more mathematical background, we could study the Einstein field equations describing how matter generates a gravitational field and how those fields react to influence the space-time trajectories of matter. We could develop an understanding of the paths, the geodesics, that particles travel in a given gravitational field and derive the tiny perihelion precession of Mercury, the fact that planetary orbits are not quite closed cyclic curves (ellipses) in Einstein's world. We could free ourselves from the restriction of just considering weak gravitational fields and consider space-time in the vicinity of black holes, where the gravitational fields are strong enough to trap light and any other bits of passing debris. We could study topics in relativistic astrophysics, model dense stars, and consider the time history of the galaxies. We could try to quantize gravity, as has been done so successfully for other classical theories such as electricity and magnetism. We could marvel at Steven Hawking's insights into the fundamental laws of black holes and quantum physics and information theory in such extreme environments. We could look for gravity's place among the other known forces of the universe and understand the insights that superstring theory provides. There is much to learn, many observations and experiments to make, and much wisdom to be found in the quest.

► **Problems** ►

7-1. A sodium lamp emits light in its rest frame with a wavelength of $5,890$ Å. If the lamp is placed on a turntable and is rotating at a speed of $0.2c$, what wavelength would an observer fixed at the center of the turntable measure?

7-2. Calculate the gravitational red-shift for a spectral line emitted at rest on the surface of the Sun and subsequently detected on Earth.

7-3. Identical atomic clocks are placed on two Boeing 747s that circle the globe at the equator, one traveling west and the other traveling east. The planes fly at an altitude of 10 km and at a speed of 0.24 km/s. After their trips around the world, the clocks are compared with one another as well as with a clock that remained behind at the airport. (The rotational speed of the surface of Earth is approximately 0.5 km/s.)

(a) What are the differences in the readings of the three clocks predicted by special relativity?

(b) What are the differences in the readings of the three clocks predicted by the gravitational red-shift?

(c) Combine the results of parts (a) and (b) to find the actual differences in the clocks' readings.

Chapter 8

Illustrations, Problems, and Discussions in General Relativity

8.1 An Aging Astronaut

When an astronaut circles the globe on the space shuttle, does she age faster or slower than we stay-at-homes on Earth?

There are two effects to consider. First, she is moving in an Earth orbit, circling Earth every 90 minutes or so. Therefore, special relativity, time dilation in particular, predicts that we measure her clocks as running slowly. Second, she resides at a higher altitude where the gravitational potential, $V(r) = -GM/r$, is higher, so general relativity predicts that her clocks run faster than ours. These two effects compete. Which wins out?

First, her velocity in a low Earth orbit is rather modest on the scale of the speed of light c. The orbit is almost circular with a radius r, and Newtonian kinematics and dynamics apply quite accurately. The centrifugal acceleration outward balances the gravitational acceleration inward, so

$$\frac{v^2}{r} = \frac{GM}{r^2},$$

where v is the astronaut's velocity and M is the mass of Earth. It is more convenient to write this in terms of the acceleration of gravity g, 9.8 m/s², observed on the surface of Earth,

$$g \equiv \frac{GM}{r_o^2},$$

where r_o is the radius of Earth. So,

$$v^2 = \frac{gr_o^2}{r}.$$

Substituting in numbers for the low Earth orbit, we find that $v \approx 7700$ m/s, so $v^2/c^2 \approx 6.6 \cdot 10^{-10}$ and all relativistic effects will be very tiny—but measurable by modern techniques. Comparing the astronaut's proper time τ to the time that passes on Earth t, taking just the relative velocity into account, gives

$$t = \gamma\tau = \frac{\tau}{\sqrt{1 - v^2/c^2}} \approx \left(1 + \frac{v^2}{2c^2}\right)\tau.$$

So,

$$\frac{\Delta t}{\tau} \approx \frac{v^2}{2c^2} = \frac{gr_o^2}{c^2 r} \tag{8.1.1a}$$

is the extra time that passes on clocks on the surface of Earth when a time τ passes on the astronaut's wristwatch.

Now we need the contribution to $\Delta t/\tau$ due to the gravitational potential difference between the astronaut and we stay-at-homes. This is given by our red-shift formula, Eq. (7.4.1b),

$$\frac{\Delta t}{\tau} \approx \frac{V(r_o) - V(r)}{c^2} = \frac{(-GM/r_o + GM/r)}{c^2}.$$

So,

$$\frac{\Delta t}{\tau} \approx \frac{gr_o^2}{c^2}(1/r - 1/r_o). \tag{8.1.1b}$$

Because r is always greater than r_o, Δt is negative, as expected—general relativity causes clocks at lower potentials to run slower, so the clock on Earth falls behind.

Finally, adding Eq. (8.1.1a) and Eq. (8.1.1b), we get our full answer,

$$\frac{\Delta t}{\tau} \approx \frac{gr_o^2}{c^2}\left(\frac{3r_o}{2r} - 1\right) \approx 7 \cdot 10^{-10}\left(\frac{3r_o}{2r} - 1\right). \tag{8.1.2}$$

What a funny result! We can enhance or reverse the relative aging processes by adjusting the height of the astronaut's orbit—if the orbit is high ($r > 3r_o/2$), the gravitational effects win and a person on the surface ages more slowly than the astronaut, but if the orbit is low ($r < 3r_o/2$), the velocity effects win and a person on the surface ages faster than the astronaut. Of course, these effects are truly tiny for Earthly conditions. But they are observable using atomic clocks, which have accuracies greater than one part in 10^{10}.

8.2 Geometry and Gravity

We have briefly mentioned in the book that the language of general relativity is differential geometry, the study of metric spaces with intrinsic curvature. Both rotating reference frames and the Schwarzschild metric illustrate this point in an elementary fashion.

For a two-dimensional surface, such as a sphere, embedded in ordinary three-dimensional flat space, we illustrate the notation of intrinsic curvature in Section 7.3. There we construct a circle on the surface of the sphere of radius R and compare its circumference C to its geodesic radius a on the sphere itself. The curvature of the sphere simply causes C to be less than $2\pi a$. We use these intrinsic quantities as ingredients into a formula

$$K = \frac{3}{\pi} \lim_{a \to 0} \left(\frac{2\pi a - C}{a^3} \right) \tag{8.2.1}$$

for the curvature of the sphere and find $K = 1/R^2$. So, smaller spheres have greater curvature that larger ones, an intuitively appealing answer.

In the four-dimensional space-time of general relativity, there is a generalization of curvature. Instead of one number K characterizing the curvature at a point, we need a whole collection of numbers, called the curvature tensor, because the curvature in space-time depends on the direction in which we make measurements. But the same simple geometric ideas apply in this higher-dimensional world, as we illustrate here.

Consider a rotating coordinate system and consider a circle of radius r whose center coincides with that of the turntable. In the nonrotating inertial frame, the circumference of the circle, C, is clearly $2\pi r$. But what does an observer at rest at radius r in the rotating reference frame measure?

Let us look at this problem in two different ways. When the observer at radius r' at rest in the rotating frame measures his circumference C', she places her meter stick down on the rotating turntable, making sure that it is perpendicular to the radius r'. An observer at rest in the nonrotating, inertial frame observes Lorentz contraction of this meter stick by a factor of $\gamma^{-1} = \sqrt{1 - v^2/c^2} = \sqrt{1 - \omega^2 r^2/c^2}$ and concludes that the number of meter sticks the rotating observer must place to go all the way around the circumference is $\gamma 2\pi r$. Both observers agree that $r' = r$ because the velocity of rotation is everywhere perpendicular to the spokes of the turntable. So, the space on the turntable is not flat, because the circumference C' is not 2π times the radius r'. We can calculate the curvature K using Eq. (8.2.1) and find for $v^2/c^2 \ll 1$ that $K = -3v^2/c^2 r^2$. Because $v = \omega r$ for the rotating turntable, we have the curvature at $r = 0$, $K = -3\omega^2/c^2$. Space on the rotating frame has a negative curvature because $C' > 2\pi r'$.

A second view of this effect comes from the metric, Eq. (7.2.6), for the turntable:

$$ds^2 = (c^2 - \omega^2 r^2)dt'^2 - \left(dr^2 + \frac{c^2 r^2 d\varphi'^2}{c^2 - \omega^2 r^2} + dz^2\right). \qquad (7.2.6)$$

Take a meter stick pointing in the transverse direction at radius r, $ds^2 = -c^2 r^2 \, d\varphi'^2/(c^2 - \omega^2 r^2)$ and compare that to a meter stick pointing in the same direction in a locally inertial frame, $ds^2 = -r^2 d\varphi^2$. We have $d\varphi = d\varphi'/\sqrt{1 - \omega^2 r^2/c^2}$, which is Lorentz contraction again because $rd\varphi$ is the proper length of the stick. Letting φ range from 0 to 2π, we see that φ' changes by less, $\Delta\varphi' = 2\pi\sqrt{1 - \omega^2 r^2/c^2}$, and we have another indication of curvature.

The Schwarzschild metric, Eq. (7.7.5), also describes a non-Euclidean space-time. For weak gravitational fields, the invariant interval reads

$$ds^2 \approx \left(1 - \frac{2GM}{c^2 r}\right)c^2 dt^2 - \left(1 + \frac{2GM}{c^2 r}\right)dr^2 - r^2(d\theta^2 + \sin^2\theta d\varphi^2). \quad (7.7.5)$$

We cannot make r small in this expression because we will violate our weak potential condition, $2GM/c^2 r \ll 1$. If we place a short rod in the radial direction from the origin, then Eq. (7.7.5) states that it has a proper length $dl \approx (1 + 2GM/c^2 r)dr$. So, if we place a meter stick, $dl = 1$ m, on a spoke from the origin, its coordinate length dr is less than 1 m and its coordinate extent is approximately $(1 - 2GM/c^2 r)$ m. This means that it takes more meter sticks to reach from r_1 to r_2 than in flat space. On the other hand, if the meter stick is placed in the transverse direction, then its proper length satisfies $dl^2 = r^2 d\theta^2$, gravity plays no role, and the circumference is given by $C = 2\pi r$. Clearly this space is also non-Euclidean. In flat space, if we increase the radius of a circle by Δl, its circumference increases by $\Delta C = 2\pi\Delta l$. But outside the mass M, where $GM/c^2 r$ is small compared to unity, when we increase the radius of a circle by a proper length Δl, the radial coordinate increases by $(1 - GM/c^2 r)\Delta l$ and the circumference, a proper distance, increases by $2\pi(1 - GM/c^2 r)\Delta l$. We again have a deficit and an indication of positive curvature.

In general relativity, particles move on trajectories determined by the curvature of space-time. Recall that in special relativity particles travel on straight lines with a constant velocity as long as no external forces are applied. These trajectories are clearly geodesics, lines of minimal length, in Minkowski space. Our simplest example of a gravitational field was provided by the acceleration of a rotating reference frame, a turntable with angular velocity ω. From the perspective of the turntable, the particle's force-free motion, which is a simple straight line in the inertial frame, becomes rather intricate when referred to the coordinates of the rotating frame. However, this complexity just represents a change of variables, discussed in detail in Section 7.2—the physics of ds^2, the particle's motion, is as simple as ever and ds^2 is the same in both descriptions. The particle travels on a geodesic between its initial and final events. This language is frame independent, so an observer fixed on the turntable agrees that the particle moves on a geodesic. When he expresses this fact mathematically in the coordinates fixed in the rotating frame, the equations are complicated—they have centripetal and Coriolis accelerations in them—but they describe the same simple physical situation, straight line motion in the inertial frame.

But Einstein argued that any gravitational field shares these properties with the example of a rotating reference frame. Gravitational accelerations are apparent just like centripetal and Coriolis accelerations. This parallelism between the two problems is precise because of the Equivalence Principle. The inertial frame in which the turntable rotates is replaced by a freely falling inertial frame in the traditional gravitational problem where we want to understand how masses move in a given environment of other masses. From the perspective of the freely falling frame, masses move in straight lines with constant velocities and light travels at the speed limit c. This situation is somewhat more intricate than the turntable problem because freely falling frames must be defined locally if the gravitational potential varies from point to point. This spatial dependence is described quantitatively through differential geometry. This formalism allows us to write the coordinate transformation from locally free falling frames to coordinates fixed relative to the mass (masses) that generate the gravitational potential. But the invariant interval ds^2 is numerically unaffected by these changes in coordinate descriptions. So, because light propagation satisfies $ds^2 = 0$ in freely falling frames, it satisfies the same equation of motion in the coordinate system at rest with respect to the masses that generate the gravitational field. In addition, massive particles that travel on straight lines (geodesics) in freely falling frames must also travel on geodesics when their motion is described relative to the coordinate system at rest relative to the masses that generate the gravitational field. The motion of particles in a spatially dependent gravitational field looks complicated (it includes, for example, planetary motion) but the space-time trajectories are just geodesics of invariant intervals such as Eqs. (7.7.2) and (7.7.5), which describe space-time in the vicinity of a given mass. The problem of gravity becomes a problem in geometry.

Let us look at a few examples of geodesics and curvature in the visualizable cases of surfaces embedded in three-dimensional Euclidean space to get some experience in this domain. In flat space on a plane, geodesics (straight lines) that are parallel in one region remain parallel everywhere, as shown in Figure 8.1a. In a curved space, such as the surface of a sphere (Figure 8.1b), this is no longer the case. Geodesics are great circles. Consider two great cir-

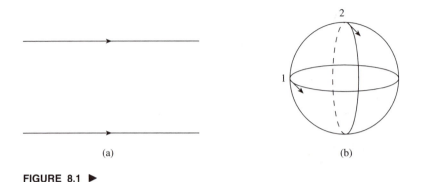

(a)   (b)

FIGURE 8.1 ▶

cles, one that coincides with the equator and another which passes through the north and south poles. Orient the great circles so that there is a point on the circle through the equator where the tangent vector 1 is parallel to a tangent vector 2 of the great circle passing through the poles, as shown in Figure 8.1b. These two geodesics, which are parallel at points 1 and 2, intersect later at the equator, separate, meet later on the back of the sphere, and so on. The surface of the sphere is curved and Euclid's theorem on parallelism does not apply. Particles constrained to move on the sphere without friction move along great circles and display the non-Euclidean geometry of the sphere. There are no real forces in this example, just the curvature of the sphere.

Analogous phenomena in the four-dimensional space-time of general relativity are the motion of the planets around the Sun and the bending of light in a gravitational field. We saw in detail in Section 7.8 that the bending of light in the vicinity of the Sun is a problem in four-dimensional space-time geometry—the light ray travels on a geodesic with $ds^2 = 0$, which corresponds to a slightly curved path as measured by our spatial coordinates (x, y, z). Similarly, the motion of the planets around the Sun is a problem in finding the geodesics having nonvanishing invariant intervals ds^2 in the Schwarzschild metric. Recall that in the context of Newton's world, the trajectories of the planets are closed ellipses. The remarkable, very special feature of these solutions is the fact that they are closed periodic figures. In other words, if a planet has a position **r** and velocity **v** at time t, then in Newton's world 1 year later the planet has exactly the same position **r** and velocity **v**. This is a special property of Newton's gravitational field, $V(r) = -GM/r$, and Newtonian space-time. If the potential is not exactly a $1/r$ form, the orbits are not closed figures. Because these Newtonian statements are surely not quite true in Einstein's world, it is not too surprising to learn that the orbits of the planets, the geodesics of the Schwarzschild metric, are not exactly closed—after 1 year passes the planets' positions and

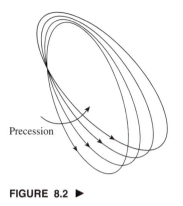

FIGURE 8.2 ▶

velocities are slightly different from the year before. This effect causes the elliptical orbits to precess, as illustrated in Figure 8.2.

This effect is truly tiny, varies as $1/c^2$, and accounts for the 43 seconds of arc precession per century (!) observed in the case of the planet Mercury. This slight failure of Newtonian mechanics was known for over a century before Einstein's new version of gravity and space-time laid it to rest.

8.3 Does Gravity Make Light Go Faster?

Although light propagates at the speed limit c in an inertial frame, it need not do so in the space near a given mass M. There are a set of experiments that have investigated this point, called radar echoes. The experimenters shot light from Earth to a nearby planet near the Sun and measured the distance to the planet as well as the time of transit of the light pulse. They then compared the total transit time of the light ray to the distance divided by the speed limit c to investigate whether gravity speeds up or slows down the light ray.

Let us work out the prediction of General Relativity for these experiments [12]. Suppose the Earth resides at the position (r_e, θ, φ) and a beam of light is transmitted to a planet at (r_p, θ, φ) and back, as shown in Figure 8.3. The nearby huge mass of the Sun creates a gravitational field that affects the trajectory of light. The gravitational fields produced by the planets are small and negligible by comparison.

First we need to calculate the physical, proper distance between Earth and the reflecting planet. This distance is not $r_e - r_p$, the radial coordinate difference, because of the gravitational field itself. In fact, from the Schwarzschild metric, the proper distance dl is related to the coordinate

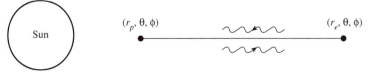

FIGURE 8.3 ▶

difference by, $dl = dr/\sqrt{1-2GM/rc^2}$. So, if the coordinate difference is $r_e - r_p$, then the physical, proper distance is

$$\int_{r_p}^{r_e} \frac{dr}{\sqrt{1-2GM/rc^2}} = \left[\sqrt{r(r-2GM/c^2)} \right.$$
$$\left. + \frac{2GM}{c^2} \ln\left(\sqrt{r} + \sqrt{r-2GM/c^2} \right) \right]_{r_p}^{r_e}.$$

Expanding the right-hand side of this result in powers of the small quantity $2GM/rc^2$, we find

$$\int_{r_p}^{r_e} \frac{dr}{\sqrt{1-2GM/rc^2}} \approx r_e - r_p + \frac{GM}{c^2} \ln\left(\frac{r_e}{r_p} \right). \qquad (8.3.1)$$

We see that the physical distance is slightly larger than the coordinate difference $r_e - r_p$.

Finally, when the light propagates along a spoke of the spherical coordinate system, variable r but fixed θ and φ, we set $ds^2 = 0$ in the Schwarzschild metric and find the relation between dr and dt,

$$\left(1 - \frac{2GM}{rc^2} \right) c^2 dt^2 = \left(1 - \frac{2GM}{rc^2} \right)^{-1} dr^2,$$

which gives

$$\frac{dr}{dt} = \pm c \left(1 - \frac{2GM}{rc^2} \right).$$

So, the coordinate time needed for the whole trip, to and fro, is

$$\Delta t = -\frac{1}{c} \int_{r_e}^{r_p} \left(1 - \frac{2GM}{rc^2} \right)^{-1} dr + \frac{1}{c} \int_{r_p}^{r_e} \left(1 - \frac{2GM}{rc^2} \right)^{-1} dr$$

$$\Delta t = \frac{2}{c} \int_{r_p}^{r_e} \left(1 - \frac{2GM}{rc^2} \right)^{-1} dr.$$

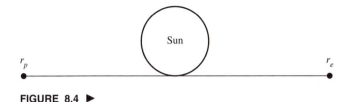

FIGURE 8.4 ▶

The proper time $\Delta\tau$ that passes at r_e where the measurements are made is related to Δt, referring to the Schwarzschild metric,

$$\Delta\tau = \sqrt{1 - \frac{2GM}{r_e c^2}}\,\Delta t$$

$$\Delta\tau = \frac{2}{c}\sqrt{1 - \frac{2GM}{r_e c^2}} \int_{r_p}^{r_e} \left(1 - \frac{2GM}{rc^2}\right)^{-1} dr$$

$$\Delta\tau = \frac{2}{c}\sqrt{1 - \frac{2GM}{r_e c^2}} \left[r_e - r_p + \frac{2GM}{c^2} \ln\left(\frac{r_e - 2GM/c^2}{r_p - 2GM/c^2}\right) \right]$$

$$\Delta\tau \approx \frac{2}{c}\left[r_e - r_p - \frac{GM\,(r_e - r_p)}{c^2}\,\frac{}{r_e} + \frac{2GM}{c^2} \ln\left(\frac{r_e}{r_p}\right) \right].$$

(8.3.2)

The difference between Eq. (8.3.1) multiplied by $2/c$ and Eq. (8.3.2) tells us how much the gravitational field influences the time of flight of the light in physical units:

$$\Delta \approx \frac{2GM}{c^3}\left[\ln\left(\frac{r_e}{r_p}\right) - \frac{(r_e - r_p)}{r_e} \right].$$

So, if $r_e \gg r_p$, then Δ is a positive quantity and the gravitational field has slowed down the light ray compared to naive expectations. The effect, however, is tiny and is very difficult to measure.

Clever experimentalists, however, applied similar calculational methods to the case where the planet lies on the other side of the Sun, as shown in Figure 8.4. This configuration produces a much larger effect because the light ray must graze by the Sun, where the gravitational field is relatively large. The prediction of general relativity was confirmed to several percent in this case.

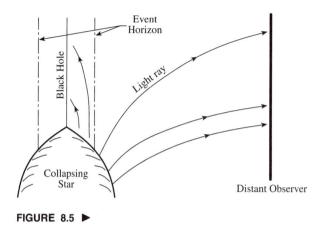

FIGURE 8.5 ▶

8.4 | Trapping Light (and Other Stuff) in Black Holes

We mentioned in our discussion of the invariant interval outside a mass M that the exact Schwarzschild metric reads

$$ds^2 = \left(1 - \frac{2GM}{c^2 r}\right) c^2 dt^2 - \frac{dr^2}{(1 - 2GM/c^2 r)} - r^2 (d\theta^2 + \sin^2 \theta \, d\varphi^2). \quad (7.7.5)$$

There is an ominous singularity in this expression, for $2GM/rc^2 = 1$, the proper time, $d\tau = \sqrt{1 - 2GM/rc^2} dt$, vanishes, and the proper distance, $dl = dr/\sqrt{1 - 2GM/rc^2}$, diverges. The critical distance,

$$r_{\text{Sch}} = \frac{2GM}{c^2},$$

is called the Schwarzschild radius. For a typical celestial body such as the Sun, r_{Sch} is much smaller than the body's actual radius and these singularities are not physical—the expression for ds^2, Eq. (7.7.5), is only true outside the massive body, just like Newton's gravitational potential, $V(r) = -GM/r$, is only true outside the mass M. However, there are astrophysical bodies that have collapsed under their own enormous weight and whose radii are less than their Schwarzschild radii. They are called black holes, and they provide a physical realization of the peculiar, extreme conditions described by the Schwarzschild metric for strong fields. The collapse of a star and the space-time paths of light rays, emitted from both inside and outside the Schwarzschild radius, are shown in Figure 8.5.

Consider a black hole and suppose that we place a clock, an emitter of light with frequency ν_e, at radius r. If we reduce r closer and closer to r_{Sch}, the clock, as measured by an observer who is far away where the gravitational field is negligible, runs slower and slower, as we can read from Eq. (7.7.5). The red-shift becomes infinite in the limit that r coincides with r_{Sch}. In effect, the gravitational field becomes so strong at r_{Sch} that light cannot escape the black hole!

The expression for the metric, Eq. (7.7.5), is not adequate for considering distances less than the Schwarzschild radius because of the singularity in the second term, $dr^2/(1 - 2GM/c^2r)$. This is a limitation of the polar coordinates used in Eq. (7.7.5) and is not a real, physical problem. By transforming to another coordinate system, the Eddington-Finklestein system in particular, we can discuss the full range of radial coordinates. Let us just review the results of this analysis because strong gravitational fields are beyond our scope, but, hopefully, not beyond the ambitions of the reader! We find that light that is emitted inside the Schwarzschild radius cannot escape—the gravitational acceleration is just too much and the light is pulled to $r = 0$. This phenomenon is the origin of the term black hole. Light that is emitted outside the Schwarzschild radius can escape, although the red-shift is considerable indeed. If a spaceship starts at $r > r_{Sch}$ but powers or falls toward the black hole, then, as measured on the spaceship clock, it takes a finite amount of time for the spaceship to pass through the Schwarzschild radius. However, a distant observer never sees the spaceship pass through the Schwarzschild radius—the spaceship disappears from view as the transmissions become increasingly red-shifted as r approaches r_{Sch}. Once the spaceship passes through r_{Sch}, communication is broken off with the outside world. Because a distant observer cannot detect events that occur inside r_{Sch}, the surface swept out in space-time by r_{Sch} is called the "event horizon". Once the spaceship is inside r_{Sch}, it moves inexorably toward smaller r. As it approaches $r = 0$, tidal forces rip it apart.

There are several candidates for black holes among the celestial bodies presently under observation. Astrophysicists predict that they are created by the gravitational collapse of very massive stars that run out of fuel late in their life cycles.

The interested student can learn more about black holes in a book [13] written at the same level as this introduction. The book also has a full bibliography and looks at recent astrophysical observations and data as well.

The modern study of black holes, their relation to quantum physics, and unification, a field pioneered by Stephen Hawking, is very stimulating because it places all of our physical laws in an extreme environment and challenges their self-consistency and our understanding of them. When you learn the elements of quantum mechanics and reconsider the properties of black holes, you will learn about Hawking radiation, the fact that black holes are, in fact hot and radiate electromagnetic waves. They are not really black at all! The unification of gravity and quantum mechanics is an active field of research to which the researchers of the high energy physics community contribute through their investigations in super-strings, a framework for a theory of everything—electricity and magnetism, radioactivity, nuclear physics, and gravity.

► Appendix A

Handy Approximations and Expansions

There is one approximation we use again and again in this book. In its simplest form, it reads

$$\frac{1}{1-x} \approx 1 + x + O(x^2). \tag{A.1}$$

This equation is useful when $x \ll 1$ so that the second order correction, denoted $O(x^2)$ in Eq. (A.1), is numerically much smaller than x itself.

The simplest proof of Eq. (A.1) consists in just multiplying through by $1 - x$, noting that the product $(1 + x)(1 - x)$ is $1 - x^2$, which differs from 1 by terms of second order. Although Eq. (A.1) can be derived using differential calculus by invoking Taylor's theorem, all that power is not needed. Simple algebra is enough.

For example, let $x = 0.01$. Then Eq. (A.1) reads that $1/0.99$ is well approximated by 1.01, with an error of order 0.0001. Clearly the linear approximation that Eq. (A.1) gives makes it very handy.

The exact equality, an infinite series expansion, behind Eq. (A.1) is

$$\frac{1}{1-x} = 1 + x + x^2 + x^3 + x^4 + \cdots, \tag{A.2}$$

where $|x| < 1$ to guarantee convergence. Another linear approximation that we use in this book, especially when we face boosts with $v/c \ll 1$, is

$$\sqrt{1-x} \approx 1 - \frac{x}{2} + O(x^2). \tag{A.3}$$

To prove this one, just square both sides.

141

Using these bits of algebra, the approximations used in the book for γ when $v/c \ll 1$ follow. For example,

$$\frac{1}{\sqrt{1 - v^2/c^2}} \approx 1 + \frac{v^2}{2c^2} + O\left(\frac{v^4}{c^4}\right). \qquad (A.4)$$

Physical Constants, Data, and Conversion Factors

1. Speed of light in empty space: $c = 2.9979\ldots \cdot 10^8$ m/s

2. Gravitational constant: $G = 6.673 \cdot 10^{-11}$ m^3/kg-s^2

3. Planck's constant: $h = 6.626 \cdot 10^{-31}$ kg-m^2/s

4. Electron's charge: $e = 1.602 \cdot 10^{-19}$ coulombs

5. Electron's rest mass: $m = 9.109 \cdot 10^{-31}$ kg $= 0.511$ MeV/c^2

6. Proton's rest mass: $m = 1.673 \cdot 10^{-27}$ kg $= 938$ MeV/c^2

7. Mass of Earth: $M = 5.9742 \cdot 10^{24}$ kg

8. Radius of Earth: $R = 6.371 \cdot 10^6$ m

9. Mass of the Sun: $M = 1.989 \cdot 10^{30}$ kg

10. Radius of the Sun: $R = 6.960 \cdot 10^8$ m

11. Distance of Earth to the Sun: $D = 1.50 \cdot 10^{11}$ m

12. Orbital speed of Earth around the Sun: $v = 2.98 \cdot 10^4$ m/s

13. Electronvolt conversion: 1 eV $= 1.602 \cdot 10^{-19}$ J

14. Kilometer conversion: 1 km $= 0.6214$ miles

15. Length of a year: 1 year $= 3.156 \cdot 10^7$ s

Selected Solutions to Problems

C.1 Chapter 2 Problems

Problem 2-1

(a) In the spaceship frame, events 1 and 2 do not occur at the same space point, that is, event 2 occurs on Earth. However, both events 1 and 2 occur at the same place in the Earth frame, so it is a proper time interval in the Earth frame.

(b) Following the same reasoning as in part (a), the time interval between events 2 and 3 is not a proper time interval in either frame.

(c) The time interval between events 1 and 3 is a proper time interval in the spaceship frame, but not in the Earth frame.

(d) Because the time between events 1 and 2 is proper time interval in the Earth frame, all that the spaceship sees is a dilated time value,

$$t_2' = \gamma t_e = \frac{10}{\sqrt{1 - \frac{v^2}{c^2}}} \, \text{min} = 12.5 \, \text{min}.$$

(e) The velocity of Earth according to the spaceship is $0.6c$. Also, the time between events 2 and 1, according to the spaceship is 12.5 minutes, as found in part (d). So the distance of Earth at event 2, according to the spaceship is

$$l_2' = 12.5 \cdot 60 \cdot 0.6c = 1.35 \cdot 10^{11} \, \text{m}.$$

(f) The time between events 3 and 2 is (according to the spaceship),

$$t_3' - t_2' = \frac{l_2'}{c} = 7.5\,\text{min}.$$

And we know the time of event 2 according to the spaceship. So the time of event 3 is

$$t_3' = 7.5 + 12.5 = 20\,\text{min}.$$

(g) From Earth's perspective, when Earth emits the pulse (event 2), the spaceship is at a distance

$$l_2 = 10\,\text{min} \cdot 0.6c = 1.08 \cdot 10^{11}\,\text{m}.$$

When the pulse reaches the spaceship, the spaceship has moved an additional distance. Let the time for the pulse to travel to the spaceship be called Δt, where

$$\Delta t = t_3 - t_2$$

and

$$c\Delta t = 1.08 \cdot 10^{11} + v\Delta t$$
$$\Delta t = \frac{1.08 \cdot 10^{11}}{2.9979 \cdot 10^8 \cdot 0.4} = 900\,\text{s} = 15\,\text{min}.$$

So the time of event 3 according to Earth is

$$t_3 = t_2 + \Delta t = 25\,\text{min}.$$

(h) We know that the time interval between events 3 and 1 is a proper time in the spaceship frame (part c). So the time interval between events 3 and 1 in the Earth frame should just be the dilated value of the time interval in the rocket frame:

$$t_3 - t_1 = \gamma(t_3' - t_1').$$

Now $t_1 = t_1' = 0$, so we should have

$$t_3 = \gamma t_3'.$$

Let us see if this is true: $t_3 = 25$ minutes, while $t_3' = 20$ minutes.

$$t_3 = \gamma t_3' = \frac{1}{\sqrt{1 - \frac{v^2}{c^2}}} 20 = \frac{5}{4} \cdot 20 = 25\,\text{min}.$$

Hence our results are consistent.

Problem 2-2

(a) The relative velocity v_r is the velocity of rocket B measured by A. The distance traveled in A's frame is 100 m (the length of rocket A in its own frame), and the time taken, as measured by A's clocks, is $1.5 \cdot 10^{-6}$ s. So,

$$v_r = \frac{100}{1.5 \cdot 10^{-6}} = 6.67 \cdot 10^7 \, \text{m/s}.$$

(b) The time measurement depends only on the relative velocity, and we know that the relative velocity of B with respect to A equals the relative velocity of A with respect to B. Hence the time (shown by the clocks on B) for the front end of A to pass the entire length of B is also $1.5 \cdot 10^{-6}$ s.

(c) An observer sitting in the front end of B sees a contracted length of rocket A. This length contraction is given by

$$l' = \frac{1}{\gamma} l = \sqrt{1 - (v_r/c)^2} \, l = 0.9749 \cdot 100 = 97.49 \, \text{m}.$$

The time taken is therefore

$$t = \frac{l'}{v_r} = \frac{l}{v_r \gamma} = \frac{1.5 \cdot 10^{-6}}{\gamma} = 0.9749 \cdot 1.5 \cdot 10^{-6} = 1.46 \cdot 10^{-6} \, \text{s}.$$

This time is not expected to agree with the time in part (b) because the events occur at two different points in A's frame, namely the front and back ends. In part (b) only the front end is being timed.

Problem 2-3

(a) Given

$$N(t) = N_0 2^{-t/T},$$

the time for one-third of the pions to decay (in the rest frame of the pions) is given by

$$\frac{2}{3} = 2^{-t'/T}$$

$$t' = T \frac{\ln(3/2)}{\ln(2)} = 1.05 \cdot 10^{-8} \, \text{sec}.$$

The length traveled by the pions in the lab frame is 35 m. So the pions see a contracted length in their own rest frame. This is given by

$$l' = \frac{l}{\gamma} = 35\sqrt{1 - v^2/c^2} \, \text{m}.$$

Also, we have

$$v = \frac{l'}{t'} = \frac{1}{1.05 \cdot 10^{-8}} 35\sqrt{1 - v^2/c^2}.$$

We solve the equation by squaring both sides

$$v^2 = \frac{10^{16}}{1.1025} 1225\left(1 - \frac{v^2}{c^2}\right)$$

$$v^2\left(1 + \frac{1111.11 \cdot 10^{16}}{9 \cdot 10^{16}}\right) = 1111.11 \cdot 10^{16}$$

$$v^2 = 8.93 \cdot 10^{16},$$

which gives us

$$v = 2.985 \cdot 10^8 \, \text{m/s},$$

which is just slightly less than the speed of light.

(b) $$l' = \frac{l}{\gamma} = 35\sqrt{1 - v^2/c^2} \, \text{m} = 2.27 \, \text{m}.$$

Problem 2-5

(a) According to the rocket frame, the signal travels from the nose of the rocket to the tail. This is just the length of the rocket in its own frame. Moreover, the speed of light is invariant, so the time taken in the rocket frame is given by

$$t' = \frac{l_0}{c}.$$

(b) At time $t = 0$, an observer in frame S measures the tail of the spaceship to be a distance l_0/γ to the left of A, according to Lorentz contraction. The tail of the spaceship is moving at velocity v to the right, so the time t when light reaches the tail is given by

$$ct = l_0/\gamma - vt.$$

Solving for t, we find

$$t = l_0/[c\gamma(1 + v/c)] = \frac{l_0}{c}\sqrt{\frac{1 - v/c}{1 + v/c}}.$$

(c) An observer in frame S sees a contracted length of the spaceship. So the time for the whole ship to pass point A is given by

$$t_2 = \frac{1}{v}\frac{l_0}{\gamma}.$$

C.2   Chapter 3 Problems

Problem 3-1

The events are illustrated in Figures C.1 through C.4.
   The figures can be used to read off the values:

   (i) $x' = 0.58$, $ct' = 0.58$

   (ii) $x' = -1.15$, $ct' = 2.3$

   (iii) $x = 1.7$, $ct = 1.7$

   (iv) $x = 1.15$, $ct' = 2.3$

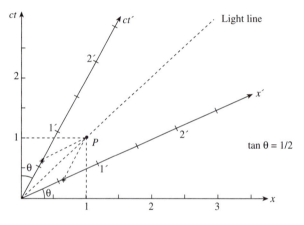

FIGURE C.1 ▶ $x = 1$, $ct = 1$.

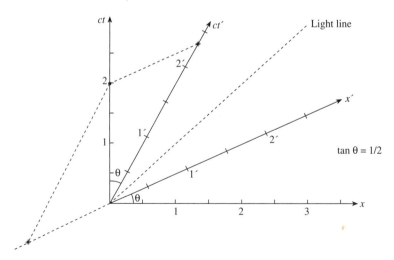

FIGURE C.2 ▶ $x = 0$, $ct = 2$.

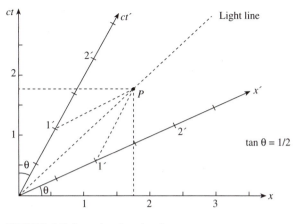

FIGURE C.3 ▶ $x' = 1$, $ct' = 1$.

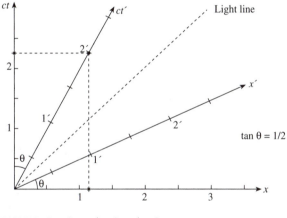

FIGURE C.4 ▶ $x' = 0$, $ct' = 2$.

Problem 3-6

The Minkowski diagram for the events is shown in Figure C.5.

From the graph, we can see that (a) the time interval between events 1 and 2 is a proper time in the Earth frame, (b) the time interval between events 2 and 3 is not a proper time interval in either frame, while (c) the time interval between events 1 and 3 is a proper time in the spaceship frame. (For an interval to be a proper time, both events should lie on the time axis of some frame.)

We also get (d) $t'_2 = 12.5$ minutes, while (e) the distance of Earth from the spaceship (as measured by the spaceship) at event 2 is $1.35 \cdot 10^{11}$ m.

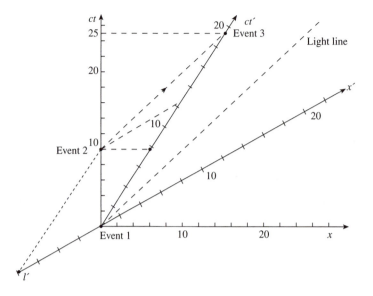

FIGURE C.5 ▶

Also (f) the time of event 3, as seen by the spaceship is 20 minutes, while (g) the time of the same event seen by Earth is 25 minutes.

Thus (h) our results from the diagram match with those of Chapter 2.

C.3 Chapter 4 Problems

Problem 4-1

Let us use invariant intervals to solve this problem. In frame S, the square of the interval is

$$(\Delta x)^2 - (c\Delta t)^2 = 0^2 - (5c)^2 = -25c^2,$$

where $\Delta x = 0$ because the events occur at the same place in S.

The value of the invariant interval is the same when measured in any inertial frame, so an observer in S' has

$$-25c^2 = (\Delta x')^2 - (c\Delta t')^2$$
$$= (\Delta x')^2 - (7c)^2$$
$$(\Delta x')^2 = 24c^2$$
$$(\Delta x') = \sqrt{24}c \text{ m},$$

which is the answer we need.

Problem 4-2

In frame S,

$$(\Delta x)^2 - (c\Delta t)^2 = (2000)^2 - 0^2 = 4 \cdot 10^6.$$

In S',

$$4 \cdot 10^6 = (\Delta x')^2 - (c\Delta t')^2$$
$$= (4000)^2 - (c\Delta t')^2$$
$$(c\Delta t')^2 = 12 \cdot 10^6$$
$$\Delta t' = \sqrt{\frac{12}{9} \cdot 10^{-10}}$$
$$= \sqrt{\frac{4}{3} \cdot 10^{-5}} \text{ s.}$$

Problem 4-3

Use the definition $\beta \equiv v/c = 4/5$ to make the notation efficient. The Lorentz transformations give

$$x = \frac{1}{\sqrt{1-\beta^2}} x' + \frac{\beta}{\sqrt{1-\beta^2}} ct'$$
$$= \frac{1}{0.6} \cdot 100 + \frac{0.8}{0.6} \cdot 3 \cdot 10^8 \cdot 9 \cdot 10^{-8}$$
$$= 202.67 \text{ m}$$

and

$$t = \frac{\beta}{\sqrt{1-\beta^2}} \frac{x'}{c} + \frac{1}{\sqrt{1-\beta^2}} t'$$
$$= \frac{0.8}{0.6} \cdot \frac{100}{3 \cdot 10^8} + \frac{1}{0.6} \cdot 9 \cdot 10^{-8}$$
$$= 59.44 \cdot 10^{-8} \text{ s.}$$

Problem 4-4

Use the Lorentz transformation,

$$t' = \frac{1}{\sqrt{1-\frac{v^2}{c^2}}} \left(t - \frac{vx}{c^2} \right).$$

(a) Equating the t' for the two events, we get

$$t_1 - \frac{vx_1}{c^2} = t_2 - \frac{vx_2}{c^2}$$

$$\frac{v}{c^2}(2L - L) = \frac{L}{2c} - \frac{L}{c}$$

$$v = -\frac{c}{2}.$$

(b)

$$t' = \frac{1}{\sqrt{3/4}}\left(\frac{L}{c} + \frac{L}{2c}\right)$$

$$= \frac{\sqrt{3}L}{c}.$$

Problem 4-5

$$\gamma = \frac{1}{\sqrt{1 - \frac{v^2}{c^2}}} = \frac{1}{\sqrt{1 - .8^2}} = \frac{5}{3}.$$

(a)

$$t' = \gamma\left(t - \frac{vx}{c^2}\right)$$

$$= \frac{5}{3}\left(5 \cdot 10^{-7} - \frac{400}{5c}\right)$$

$$= \frac{5}{3} \cdot 10^{-7}\left(5 - \frac{8}{3}\right)$$

$$= 1.48 \cdot 10^{-7} \text{ s.}$$

$$x' = \frac{5}{3}\left(100 - \frac{4c}{5} \cdot 5 \cdot 10^{-7}\right)$$

$$= -\frac{100}{3} \text{ m.}$$

(b)

$$t'_2 = \frac{5}{3}\left(7 \cdot 10^{-7} - \frac{200}{5c}\right)$$

$$= 11.67 \cdot 10^{-7} \text{ s}$$

and the time difference is

$$\Delta t' = t'_2 - t' = 10.19 \cdot 10^{-7} \text{ s.}$$

Problem 4-7

(a) To find the velocity of the particle in frame S, we use the relativistic addition formula,

$$u = \frac{0.6c - c/3}{1 - 0.6/3} = \frac{c}{3}.$$

(b) Use the invariance of the space-time interval,

$$(\Delta x)^2 - c^2(\Delta t)^2 = (\Delta x')^2 - c^2(\Delta t')^2.$$

We know $\Delta t' = 3 \cdot 10^{-7} - 10^{-7}$ s, and

$$\Delta x' = u' \Delta t' = \frac{-c}{3} \cdot 2 \cdot 10^{-7}.$$

Also we know that $\Delta t = \Delta x/u$, which gives us

$$(\Delta x)^2 \left(1 - \frac{c^2}{u^2}\right) = 4 \cdot 10^{-14} \left(\frac{c^2}{9} - c^2\right),$$

which gives, using u from part (a),

$$(\Delta x) = \frac{2c}{3} \cdot 10^{-7} = 20 \text{ m}.$$

We could have also done the problem equally easily using Lorentz transformations.

Problem 4-9

Let v_S be the velocity of light in the frame S. This is given by the relativistic addition formula

$$v_S = \frac{v_m + v}{1 + \frac{v_m v}{c^2}}.$$

We expand the denominator in a Taylor series,

$$v_S = (v_m + v)\left(1 + \frac{v_m v}{c^2}\right)^{-1}$$

$$= (v_m + v)\left(1 - \frac{v_m v}{c^2} + \cdots\right),$$

where we have expanded only to leading order in v/c. We expand the bracket, again dropping the term with v^2/c^2:

$$v_S = v_m + v\left(1 - \frac{v_m^2}{c^2}\right).$$

Problem 4-11

(a) Let Earth be at the origin and the star be somewhere in the first
quadrant (i.e., it has positive x and positive y coordinates). Then the
light signal that is emitted from the star toward Earth has negative
components. If Earth and the star did not have a relative velocity
these components would be

$$v_x = -c \cos \theta$$

$$v_y = -c \sin \theta.$$

Now suppose that the star (frame S) is moving away from Earth
(frame S′) with velocity v in the x direction. We therefore know
the x component of velocity in S, and we know the x component
of velocity of S′ with respect to S. So, to find the x component of
velocity in S′, we use the relativistic addition formula. Also, we note
that because the speed of light is invariant, the only way the x component can change is if the angle θ changes. In S′, the x component
of light is therefore given by

$$v'_x = -c \cos \theta'.$$

So,

$$-c \cos \theta' = \frac{-c \cos \theta - v}{1 - \frac{(v)(-c \cos \theta)}{c^2}}.$$

Dividing through by $-c$, we get the desired result:

$$\cos \theta' = \frac{\cos \theta + \frac{v}{c}}{1 + \frac{v \cos \theta}{c}}.$$

(b) As in Problem 4.9, we expand the denominator to first order in v/c,

$$\cos \theta' \approx \left(\cos \theta + \frac{v}{c} \right) \left(1 - \frac{v \cos \theta}{c} \right)$$

$$\cos \theta' \approx \cos \theta + \frac{v}{c} - \frac{v}{c} \cos^2 \theta$$

$$\approx \cos \theta + \frac{v}{c} \sin^2 \theta.$$

(c) We now have $\theta \approx \theta'$. We have from part (b)

$$\cos \theta' - \cos \theta = \frac{v}{c} \sin^2 \theta.$$

We use a trigonometric identity

$$\cos \theta' - \cos \theta = 2 \left(\sin \frac{\theta - \theta'}{2} \right) \left(\sin \frac{\theta + \theta'}{2} \right).$$

We substitute $\alpha = \theta' - \theta$, and use the approximation $\sin \beta \approx \beta$ for $|\beta| \ll 1$. We also use the approximate equality of θ and θ' in the second sine term in the product to get

$$-2\frac{\alpha}{2}\left(\sin \frac{2\theta}{2}\right) \approx \frac{v}{c}\sin^2 \theta$$

$$\alpha \approx -\frac{v}{c}\sin \theta.$$

Problem 4-13

(a) We just use the relativistic addition formulae

$$v_x = \frac{v'_x + v}{1 + \frac{(v'_x v)}{c^2}}$$

$$v_y = \frac{v'_y}{\gamma\left(1 + \frac{(v'_x v)}{c^2}\right)} = \frac{v'_y}{\gamma}.$$

(b) v_y and v'_y are different, even though the y coordinates remain unchanged, because time dilates.

Problem 4-14

(a) An observer at rest in frame S' measures the position of the meter stick at a particular time t' in his frame. But, by the relativity of simultaneity, a clock at $-.5$ m lags the time on a clock at $+0.5$ m when viewed from S'. Therefore, an observer in S' states that the right end of the rod reaches $y = 0$ before the left end of the rod. The rod is, therefore, seen to tilt up to the right, as shown in Figure C.6. Curious!

(b) First find the time and position in frame S' when the right end of the meter stick crosses the x axis. This event occurs in the S frame at $t_1 = 0$ and $x_1 = 0.5$ m, $y_1 = 0$. Using Lorentz transformations gives

$$x'_1 = \gamma(x_1 - vt_1) = 0.5\gamma$$

$$t'_1 = \gamma\left(t_1 - \frac{vx_1}{c^2}\right) = -\frac{0.5\gamma v}{c^2}$$

$$y'_1 = 0.$$

Now we look at another event, that of the midpoint of the meter stick crossing the x axis. In the S frame, this happens at $x_2 = y_2 = 0$, $t_2 = 0$. In the S' frame, we have

$$x'_2 = y'_2 = t'_2 = 0.$$

Denote the right end coordinates at time t_2 to be (x', y').

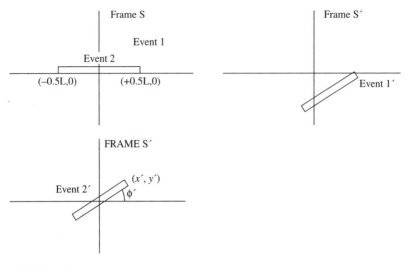

FIGURE C.6 ▶

From Problem 4.13, we compute the velocities in the x and y directions in the S′ frame, $v'_x = -v$ and $v'_y = v_y/\gamma$,

$$x' = x'_1 + v'_x(t'_2 - t'_1)$$

$$= 0.5\gamma - v\frac{0.5\gamma v}{c^2} = 0.5\gamma(1 - v^2/c^2) = \frac{0.5}{\gamma}$$

$$y' = y'1 + v'_y(t'_2 - t'_1)$$

$$= 0 + \frac{v_y}{\gamma}\frac{0.5\gamma v}{c^2} = \frac{vv_y}{c^2}0.5.$$

And we have

$$\tan\phi' = \frac{y'}{x'} = \frac{vv_y\gamma}{c^2}.$$

C.4   Chapter 5 Problems

Problem 5-2

(a) Let the distance of the observer from point 2 be x. The first pulse, therefore, has to travel a distance of

$$l_1 = d + x = vt + x$$

and the second pulse has to travel a distance of x meters. The observer, therefore, receives the second pulse at a time

$$t_2 = \frac{x}{c} + t$$

and we have

$$t_1 = \frac{l_1}{c} = \frac{x}{c} + t\frac{v}{c}.$$

So the interval between the pulses is

$$T = t_2 - t_1 = t\left(1 - \frac{v}{c}\right).$$

(b)

$$\frac{d}{T} = \frac{v}{1 - v/c}$$

follows from part (a). This is the apparent speed the observer sees.

(c) If an observer is a distance y to the left of point 1, the two pulses will be received at times

$$t_1 = \frac{y}{c}$$

and

$$t_2 = \frac{y + vt}{c} + t.$$

The time interval is, therefore,

$$T' = t\left(1 + \frac{v}{c}\right)$$

and the apparent speed is

$$\frac{d}{T'} = \frac{v}{1 + (v/c)}.$$

As v approaches c, d/T' approaches $c/2$.

C.5 **Chapter 6 Problems**

Problem 6-1

The rest mass energy of the proton is 938 MeV. We use the formula for relativistic energy to get

$$E = \frac{mc^2}{\sqrt{1 - v^2/c^2}}$$

$$10^{19}\,\text{eV} = \frac{938 \cdot 10^6\,\text{eV}}{\sqrt{1 - v^2/c^2}}.$$

Solving for v, we get

$$\gamma = \frac{10^{19}}{938 \cdot 10^6} = 1.067 \cdot 10^{10}$$

$$\frac{v}{c} = \sqrt{1 - (938 \cdot 10^{-13})^2} \approx 1 - \frac{(938 \cdot 10^{-13})^2}{2}.$$

So, the proton is traveling very close to the speed limit. The fractional difference is only $-4.40 \cdot 10^{-21}$.

(a) The time it takes to traverse the galaxy in the galaxy frame is given by

$$t_{\text{gal}} = \frac{10^5 c \text{ year}}{v} = \frac{10^5 \text{ year}}{1 - (938 \cdot 10^{-13})^2/2} \approx 10^5 \text{ years}$$

(b) The event of the proton traversing the galaxy is a proper time interval in the proton's frame because the two events, the proton starting off and the proton reaching the edge of the galaxy, occur at the same space point in the proton's frame. Therefore, the time seen by the galaxy between these two events is just the dilated proton time for the same events. But we know the galaxy time from part (a), so we have

$$t_{\text{gal}} = \gamma t_{\text{proton}}$$

$$t_{\text{proton}} = \frac{t_{\text{gal}}}{\gamma}$$

$$= \frac{10^5 \text{ year}}{1 - (938 \cdot 10^{-13})^2/2} \frac{1}{1.067 \cdot 10^{10}}$$

$$\approx 9.38 \cdot 10^{-6} \text{ years} \approx 4.93 \text{ minutes}.$$

Problem 6-2

The rest mass energy of an electron is 0.51 MeV. The energy acquired in the potential drop is 10^5 eV. So, the relativistic energy, γmc^2, is,

$$10^5 + 0.51 \cdot 10^6 = \frac{0.51 \cdot 10^6}{\sqrt{1 - \frac{v^2}{c^2}}}$$

$$\sqrt{1 - \frac{v^2}{c^2}} = \frac{5.1}{6.1}$$

$$\frac{v}{c} = 0.55.$$

(a) The time to travel 10 m in the lab frame is

$$t_{\text{lab}} = \frac{10}{v} = \frac{10}{0.55 \cdot 3 \cdot 10^8} = 6.06 \cdot 10^{-8} \text{ s.}$$

(b) Following the same reasoning as in Problem 6.1, we have Lorentz contraction,

$$d_{\text{el}} = \frac{d_{\text{lab}}}{\gamma} = \frac{5.1}{6.1} \cdot 10 \text{ m} = 8.36 \text{ m.}$$

Problem 6-4

The equations are:

(a) Newtonian kinematics,

$$E = \frac{p^2}{2m} + mc^2,$$

where we included the rest mass energy, mc^2, for easy comparison to the relativistic formulas.

(b) Relativistic kinematics for a particle with rest mass m,

$$E = +\sqrt{p^2c^2 + m^2c^4}.$$

(c) Relativistic kinematics for a massless particle (i.e., photon),

$$E = |\, p \,| \, c.$$

The graphs are shown in Figure C.7. Curves 1 and 2 agree near $E = mc^2$, whereas curves 2 and 3 agree at large momentum values.

Problem 6-5

From the Lorentz transformation, we have

$$E' = \gamma(E - vp_1).$$

If we specialize to the case of the photon, then $p_1 = E/c$, and we get

$$E' = \gamma\left(E - v\frac{E}{c}\right)$$

$$= \frac{1}{\sqrt{1 - \frac{v^2}{c^2}}} E\left(1 - \frac{v}{c}\right)$$

$$= \sqrt{\frac{1 - v/c}{1 + v/c}} E.$$

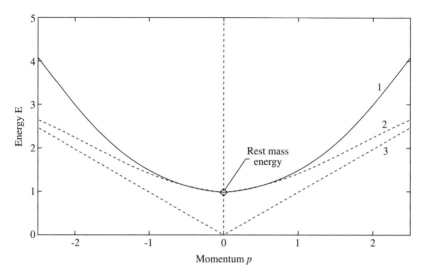

FIGURE C.7 ▶

Problem 6-9

The electrostatic potential energy is given by

$$U = \frac{3kQ^2}{5r}.$$

Here Q is the total charge, which is

$$Q = 1.6 \cdot 10^{-19} \cdot n,$$

where n is the number of electrons in 1 g of electrons:

$$n = \frac{1 \cdot 10^{-3}}{9.1 \cdot 10^{-31}} = 1.1 \cdot 10^{27}.$$

So, $Q = -1.76 \cdot 10^8$ coulombs and

$$U = \frac{3 \cdot 9 \cdot 10^9 \cdot 3.10 \cdot 10^{16}}{5 \cdot .1} = 1.67 \cdot 10^{27}.$$

Equating this to the relativistic rest mass energy, we get

$$mc^2 = U.$$

So,

$$m = 1.86 \cdot 10^{10} \text{ kg},$$

which is amazingly large!

Problem 6-10

Consider the equation requiring that the position of the center of mass of the system does not change. In view of the approximations stated in this problem, this is

$$\Delta \bar{x} = 0 = mL + M\Delta x,$$

where m is the mass equivalent of light, M the mass of the box (neglecting the loss of mass during light transit), and Δx is the amount moved by the box. Now, if we include the fact that the box loses mass during light transit, the second term becomes $(M - m)\Delta x$, and if we include the fact that the box has moved (and hence the light travels less distance before being absorbed), the first term becomes $m(L + \Delta x)$. The signs have been taken according to the conventions in the book. So our equation now reads

$$0 = m(L + \Delta x) + (M - m)\Delta x,$$

which gives us exactly the same equation as the one that was derived using the approximations.

Problem 6-11

The relativistic kinetic energy is given by $E - mc^2$, where E is the total energy

$$E - mc^2 = 2mc^2$$

$$E = 3mc^2.$$

Using the formula for relativistic energy, we get

$$E = \frac{mc^2}{\sqrt{1 - \frac{v^2}{c^2}}}$$

$$3mc^2 = \frac{mc^2}{\sqrt{1 - \frac{v^2}{c^2}}}$$

$$\sqrt{1 - \frac{v^2}{c^2}} = \frac{1}{3}$$

$$v = 0.94c.$$

The momentum is given by

$$p = \frac{mv}{\sqrt{1 - \frac{v^2}{c^2}}} = 3mv = 2.82mc.$$

If the relativistic kinetic energy is five times the rest mass then, $E = 6mc^2$, and we have

$$\sqrt{1 - \frac{v^2}{c^2}} = \frac{1}{6}$$

$$v = 0.986c$$

$$p = 6mv = 5.916mc.$$

Problem 6-12

Since $v/c = 0.99$, the particle's $\gamma = 1/\sqrt{1 - (0.99)^2} = 7.09$. For a potential drop of X:

$$X \text{ eV} = \frac{mc^2}{\sqrt{1 - \frac{v^2}{c^2}}} - mc^2 = 6.09 \cdot 0.51 \cdot 10^6 \text{ eV}$$

$$X = 3.11 \cdot 10^6 \text{ V}.$$

The proton's rest mass is 938 MeV, so the value of X becomes

$$X = 6.09 \cdot 938 \cdot 10^6 = 5712 \cdot 10^6 \text{ V}.$$

Problem 6-14

Let the incident photon energy be E_0 and the reflected energy be E_f. Let the final velocity of the rocket be v. Energy-momentum conservation reads,

$$\frac{E_0}{c} = \gamma m_0 v - \frac{E_f}{c}$$

$$E_0 + m_0 c^2 = E_f + \gamma m_0 c^2.$$

These give

$$E_0 + E_f = \gamma m_0 vc$$

$$E_0 - E_f = \gamma m_0 c^2 - m_0 c^2.$$

Thus,

$$2E_0 = \gamma m_0 vc + \gamma m_0 c^2 - m_0 c^2$$

$$E_0 = \frac{1}{2} m_0 c^2 \left[\gamma \left(1 + \frac{v}{c} \right) - 1 \right].$$

The mass equivalent is

$$\frac{E_0}{c^2} = \frac{1}{2} m_0 \left[\gamma \left(1 + \frac{v}{c} \right) - 1 \right].$$

Problem 6-16

Let the photons have a total energy E in the initial rest frame of the rocket. Energy and momentum conservation read

$$M_i c^2 = E + \gamma M_f c^2$$

$$0 = \gamma M_f v - \frac{E}{c}.$$

This gives

$$\gamma M_f v = \frac{E}{c}$$

$$M_i c = \frac{E}{c} + \gamma M_f c.$$

Putting the first equation into the second, we get

$$M_i c = \gamma M_f (c + v)$$

$$\frac{M_i}{M_f} = \frac{1 + \frac{v}{c}}{\sqrt{1 - \frac{v^2}{c^2}}}$$

$$= \left(\frac{1 + \frac{v}{c}}{1 - \frac{v}{c}} \right)^{1/2}.$$

Problem 6-17

In the rest frame of the K^o meson, the two π mesons are emitted in opposite directions, so that momentum is conserved. We are observing from the lab frame, in which the K^o meson is moving, so the problem is one of relativistic velocity addition. The speed of the meson that is moving in the same direction as the K meson in the lab frame is

$$v_{max} = \frac{v_1 + v_r}{1 + v_1 v_r / c^2}$$

$$= \frac{0.9c + 0.85c}{1 + 0.765}$$

$$= 0.992c$$

and the minimum speed is that of the other pion,

$$v_{min} = \frac{v_2 + v_r}{1 + v_2 v_r / c^2}$$

$$= \frac{0.9c - 0.85c}{1 - 0.765}$$

$$= 0.212c.$$

Problem 6-18

The velocity of B observed by A is, using relativistic addition of velocities,

$$v_r = \frac{2v}{1 + v^2/c^2}.$$

The energy is given by

$$E = \frac{M_0 c^2}{\sqrt{1 - v_r^2/c^2}}$$

$$\begin{aligned}
1 - \frac{v_r^2}{c^2} &= 1 - \frac{4v^2}{c^2(1 + v^2/c^2)^2} \\
&= \frac{(1 + v^2/c^2)^2 c^2 - 4v^2}{(1 + v^2/c^2)^2 c^2} \\
&= \frac{c^2(1 - v^2/c^2)^2}{c^2(1 + v^2/c^2)^2}.
\end{aligned}$$

So we get

$$E = M_0 c^2 \frac{1 + v^2/c^2}{1 - v^2/c^2}.$$

Problem 6-19

(a) We know that for a relativistic system, energies add, momenta add (vectorially), but rest masses do not add. So the total energy of the two photons is $E_{\text{total}} = 200 + 400 = 600$ MeV.

The momentum of a photon is given by $p = E/c$, and the total momentum of the system is

$$p_{\text{total}} = \sqrt{(400/c)^2 + (200/c)^2} = 200\sqrt{5} \text{ MeV}/c$$

at an angle $\theta = 63.43°$ above the x axis.

(b) If a single particle had the total energy and total momentum of the system, it would have a rest mass m given by

$$E_{\text{total}}^2 - p_{\text{total}}^2 c^2 = m^2 c^4$$
$$600^2 - 5 \cdot 200^2 = m^2 c^4,$$

which gives

$$m = 400 \text{ Mev}/c^2.$$

The single particle would travel along the direction making an angle $\theta = 63.43°$ above the x axis.

To find the speed, we have

$$\frac{v}{c^2} = \frac{p_{\text{total}}}{E_{\text{total}}}$$

$$\frac{v}{c} = \frac{p_{\text{total}}c}{E_{\text{total}}}$$

$$v = \frac{\sqrt{5}}{3}c.$$

Problem 6-20

The total energy is

$$E = m_o c^2 + 3m_o c^2 + 2m_o c^2 = 6m_o c^2.$$

The total momentum is carried solely by the first particle, and this is

$$p^2 c^2 = E_1^2 - m_o^2 c^4,$$

where $E_1 = m_o c^2 + 3m_o c^2 = 4m_o c^2$.

$$p^2 = 15m_o^2 c^2.$$

Now we consider the composite particle.

$$E^2 - p^2 c^2 = M_o^2 c^4.$$

This gives

$$M_o^2 = 36m_o^2 - 15m_o^2 = 21m_o^2.$$

So $M_o = \sqrt{21}m_o$.

Problem 6-21

(a) Using momentum and energy conservation relations,

$$\frac{E}{c} = \gamma m v$$

$$E + m_o c^2 = \gamma m c^2.$$

Performing the algebra, we get the velocity,

$$\frac{v}{c} = \frac{E}{\gamma m c^2} = \frac{E}{E + m_o c^2}.$$

Next we calculate $1/\gamma$ in terms of m_o and E:

$$\frac{1}{\gamma} = \sqrt{1 - \frac{v^2}{c^2}} = \frac{(2Em_oc^2 + m_o^2c^4)^{1/2}}{E + m_oc^2}.$$

Now we get m, the mass of the composite particle:

$$m = \frac{1}{\gamma}\left(m_o + \frac{E}{c^2}\right) = \sqrt{m_o^2 + \frac{2Em_o}{c^2}}.$$

(b) The incident particle has a velocity of $v_i = 4c/5$ m/s, and a γ_i given by

$$\gamma_i = (\sqrt{1 - 16/25})^{-1} = 5/3.$$

Employing the conservation relations, we have

$$\frac{5}{3} \cdot \frac{4}{5}m_oc = \frac{4}{3}m_oc = \gamma mv$$

$$\frac{5}{3}m_oc^2 + m_oc^2 = \gamma mc^2.$$

Algebra gives

$$\frac{4}{3}m_oc = \gamma mv$$

$$\frac{8}{3}m_o = \gamma m.$$

So,

$$v = \frac{c}{2}$$

$$\gamma = \frac{1}{\sqrt{1 - 1/4}} = \frac{2}{\sqrt{3}},$$

which gives

$$\frac{8}{3}m_o = \frac{2}{\sqrt{3}}m$$

$$m = \frac{4}{\sqrt{3}}m_o.$$

Problem 6-23

(a) We have the kinetic energy,

$$K = (\gamma - 1)m_o c^2.$$

The values given are $m_o = 135$ MeV/c^2, $K = 1$ GeV, so

$$10^9 = (\gamma - 1) \cdot 135 \cdot 10^6$$

$$\gamma - 1 = 1000/135$$

$$\gamma = \frac{1}{\sqrt{1 - \frac{v^2}{c^2}}} = 8.4074$$

$$v = 2.977 \cdot 10^8 \text{ m/s}.$$

Let E_1 and E_2 be the energies of the photons, where the photon with E_2 travels in the direction of the incident π meson. We have

$$\gamma m_o v = \frac{E_2}{c} - \frac{E_1}{c}$$

$$\gamma m_o c^2 = E_1 + E_2.$$

We get

$$E_2 + E_1 = \gamma m_o c^2$$

$$E_2 - E_1 = \gamma m_o cv,$$

which gives

$$E_2 = \frac{1}{2}\gamma m_o c(c+v) = 1.131 \text{ GeV}$$

$$E_1 = \frac{1}{2}\gamma m_o c(c-v) = 3.973 \text{ MeV}.$$

(b) Let θ be the angle between one photon and the incident direction. Then the angle included between the two photons is 2θ. We have

$$\gamma m_o c^2 = 2E$$

$$\gamma m_o v = \frac{2E}{c} \cos \theta.$$

So

$$\gamma m_o v = \gamma m_o c \cos \theta$$

$$\cos \theta = \frac{v}{c}$$

$$\theta = 6.769°$$

and the included angle is $2\theta = 13.539°$.

Problem 6-24

(a) We have one photon going forward and one backward. The conservation relations read

$$E_{\bar{p}} + E_p = E_1 + E_2$$
$$1000 + 2 \cdot 938 = E_1 + E_2$$

and

$$p_{\bar{p}} = \frac{1}{c}(E_1 - E_2),$$

where $p_{\bar{p}}$ can be found simply:

$$p_{\bar{p}}^2 c^2 = E_{\bar{p}}^2 - m^2 c^4$$
$$p_{\bar{p}}^2 c^2 = 1938^2 - 938^2$$
$$p_{\bar{p}} c = 1695.9 \text{ MeV.}$$

The two equations give

$$E_1 = 2286 \text{ MeV}$$
$$E_2 = 590 \text{ MeV.}$$

(b) We now find the velocity of the anti-proton,

$$1938 \text{ MeV} = \frac{938 \text{ MeV}}{\sqrt{1 - \frac{v^2}{c^2}}},$$

which gives

$$v = 0.875c.$$

So the energy of the photons measured in this frame are

$$E_1' = \gamma(E_1 - vp_1) = \gamma\left(E_1 - \frac{v}{c}E_1\right)$$

$$E_1' = \sqrt{\frac{1 - v/c}{1 + v/c}} E_1 = \sqrt{\frac{1 - .875}{1 + .875}} 2286 = 590 \text{ MeV}$$

$$E_2' = \gamma(E_2 - vp_2) = \gamma\left(E_2 + \frac{v}{c}E_2\right)$$

$$E_2' = \sqrt{\frac{1 + v/c}{1 - v/c}} E_1 = \sqrt{\frac{1 + 0.875}{1 - 0.875}} 590 = 2286 \text{ MeV.}$$

So, in this frame the energies of the photons switch—which we could have anticipated without calculation!

Problem 6-26

If Q is the energy of the photon, we have

$$Q + \gamma_o m_o c^2 = Q + \gamma_f m_o c^2$$

$$\frac{Q}{c} - \gamma_o m v_o = -\frac{Q}{c} + \gamma_f m_o v_f,$$

where v_f and γ_f, refer to the electron after the collision. The first equation gives $\gamma_o = \gamma_f$, which implies $v_o = v_f$. So

$$\frac{2Q}{c} = 2\gamma_o m_o v_o$$

$$\frac{Q}{c} = \frac{m v_o}{\sqrt{1 - v_o^2/c^2}}$$

$$v_o^2 = \frac{Q^2}{m^2 c^2}\left(1 - \frac{v_o^2}{c^2}\right)$$

$$v_o^2\left(1 + \frac{Q^2}{m_o^2 c^4}\right) = \frac{Q^2}{m_o^2 c^2}$$

$$v_o = \frac{Qc}{\sqrt{Q^2 + m_o^2 c^4}},$$

which is the required answer.

Problem 6-28

Let the momentum of the K meson be p. Because one of the created pions is at rest, the second pion's momentum must be p. The conservation relations read

$$E_K = 137 + E_\pi$$

$$p_K = p_\pi,$$

where the subscript π refers to the moving pion. Squaring the momentum relation equation, multiplying by c^2, and subtracting from the square of the energy conservation equation, we get

$$E_K^2 - p_K^2 c^2 = 137^2 + 2 \cdot 137 \cdot E_\pi + E_\pi^2 - p_\pi^2 c^2.$$

The left-hand side and the last two terms of the right-hand side are clearly the expressions for rest mass energies $m_K^2 c^4$ and $m_\pi^2 c^4$, which we know, so we get

$$494^2 = 137^2 + 274 E_\pi + 137^2 = 37538 + 274 E_\pi$$

$$E_\pi = 753.6 \text{ MeV.}$$

The energy of the original K meson is given by

$$E_K = 137 + E_\pi = 890.6 \text{ MeV}.$$

Problem 6-29

The configuration that has the minimum final energy is that in which all the three created particles move in the same direction with equal velocities. (Viewed from the center of momentum frame, this means that all the particles in the final state are created at rest there.) In such a case, the conservation relations read in the lab frame (where E is the gamma-ray energy),

$$E + 0.51 \text{ Mev} = 3E_e,$$

where E_e is the energy of each of the created particles.

$$p_e = \frac{E}{3c}$$

because momenta are shared equally. Combining these equations,

$$E_e^2 = 0.51^2 + p_e^2 c^2 = 0.51^2 + \frac{E^2}{9} = \left(\frac{E + 0.51}{3}\right)^2.$$

We can solve for E:

$$9 \cdot \left(0.51^2 + \frac{E^2}{9}\right) = E^2 + 0.51^2 + 1.02E$$

$$E = \frac{8 \cdot 0.51^2}{1.02} = 2.04 \text{ MeV}.$$

Problem 6-30

Let's use invariant methods to solve this problem. Let p be the energy-momentum four-vector $p = (E/c, \vec{p})$. The invariant length of p is $p^2 = E^2/c^2 - \vec{p}^2 = m_o^2 c^2$, where we identified the rest mass m_o in the special case that p refers to a single particle. Note that p^2 is the same in all frames, so we could evaluate it in any frame to solve a practical problem.

In this problem, energy-momentum conservation written as a four-vector relation is $p_1 + p_2 = p_3 + p_4 + p_X$ in an obvious notation. Because the four-vectors $p_1 + p_2$ and $p_3 + p_4 + p_X$ are equal, so are their invariant lengths,

$$(p_1 + p_2)^2 = (p_3 + p_4 + p_X)^2.$$

Evaluate the left-hand side of this equation in the lab frame and the right-hand side in the center of momentum frame. To obtain the maximum rest mass m_X, all three final state particles should be produced at rest in the center of momentum frame. So,

$$c^2(p_3 + p_4 + p_X)^2 = (m_o c^2 + m_o c^2 + m_X c^2)^2.$$

The left-hand side is easily evaluated in terms of lab quantities,

$$c^2(p_1 + p_2)^2 = (E_1 + m_o c^2)^2 - c^2 \vec{p}^2 = m_o^2 c^4 + 2m_o c^2 E_1 + E_1^2 - \vec{p}_1^2 c^2.$$

But $E_1^2 - \vec{p}_1^2 c^2 = m_o^2 c^4$, so,

$$c^2(p_1 + p_2)^2 = 2m_o^2 c^4 + 2m_o c^2 E_1.$$

Combining these two results,

$$2m_o^2 c^4 + 2m_o c^2 E_1 = (2m_o c^2 + m_X c^2)^2,$$

which gives,

$$m_X c^2 = \sqrt{2m_o c^2 (m_o c^2 + E_1)} - 2m_o c^2 = 219 \text{ GeV}$$

after substituting in $m_o c^2 = 0.938$ GeV/c^2, and $E_1 = 300$ GeV $+ 0.938$ GeV.

A nice aspect of this method of solution is that we did not have to calculate any velocities in intermediate steps on the way to determining m_X.

Problem 6-31

(a) One photon emerges with energy E_1 at right angles to the incident line of motion. Let the other photon emerge with energy E_2 at an angle θ from the incident line of motion. The total initial energy (in MeV) is 0.511 (rest mass of positron) + 0.511 (kinetic energy of positron) + 0.511 (rest mass of electron). So we get

$$3 \cdot 0.511 = E_1 + E_2.$$

The initial momentum is due to the positron and is in the incident direction. Its value is

$$pc = \sqrt{E^2 - m^2 c^4} = \sqrt{(0.511 \cdot 2)^2 - 0.511^2} = 0.511\sqrt{3} \text{ MeV}.$$

The momentum conservation in the right-angle direction to the incident line of motion is

$$\frac{E_1}{c} = \frac{E_2}{c} \sin \theta$$

and in the incident direction

$$\frac{E_2}{c} \cos \theta = \frac{0.511\sqrt{3}}{c}.$$

Squaring and adding the last two relations $(\sin^2 \theta + \cos^2 \theta = 1)$, we get

$$E_2^2 = 0.511^2 \cdot 3 + E_1^2.$$

Squaring the energy conservation equation, we get

$$E_2^2 = 9 \cdot 0.511^2 - 6 \cdot 0.511 \cdot E_1 + E_1^2.$$

Equating the last two equations, we get

$$E_1 = 0.511 \text{ MeV}.$$

Energy conservation gives

$$E_2 = 3 \cdot 0.511 - E_1 = 1.022 \text{ MeV}.$$

(b) The angle is found from the momentum conservation relation,

$$E_1 = E_2 \sin \theta$$

$$\sin \theta = \frac{E_1}{E_2},$$

which gives $\theta = 30°$.

Problem 6-33

(a) The binding energy is found from the definition given in the problem:

$$\frac{1}{c^2} \Delta E = m_{Be} - 4m_p - 3m_n$$

$$= 6536 - 4 \cdot 938.28 - 3 \cdot 939.57 = -35.83 \text{ MeV}/c^2.$$

(b) The initial energy is $E_{Be} + E_n = 6536 + 939.57 = 7475.57$ MeV. The energy is split up equally between the alpha particles, so

$$7475.57 = 2(K + mc^2) = 2(K + 3728),$$

which gives

$$K = 9.785 \text{ MeV}.$$

References

[1] A. P. French, *Special Relativity*, W. W. Norton, New York, 1968.

[2] N. D. Mermin, *Space Time in Special Relativity*, Waveland Press, Prospect Heights, IL, 1968.

[3] E. F. Taylor and J. A. Wheeler, *Spacetime Physics*, W. H. Freeman, New York, 1992.

[4] W. Rindler, *Introduction to Special Relativity*, Oxford University Press, Oxford, 1991.

[5] W. Rindler, *Essential Relativity*, Springer-Verlag, Berlin, 1971.

[6] C. G. Darwin, "The Clock Paradox in Relativity," *Nature* 180, 976 (1957).

[7] M. Born, *Einstein's Theory of Relativity*, Dover Publications, New York, 1962.

[8] R. P. Feynman, *Photon-Hadron Interactions*, W. A. Benjamin, Inc., MA, 1972.

[9] R. V. Pound and G. A. Rebka, Jr., "Apparent Weight of Photons," *Physics Rev. Lett.* 4, 337 (1960); R. V. Pound and J. L. Snider, "Effect of Gravity on Gamma Radiation," *Physics Rev.* B140, 788 (1965).

[10] O. R. Frisch, "Time and Relativity," *Contemporary Physics* 3, 16 (1961); 3, 194 (1962).

[11] R. Adler, M. Bazin, and M. Schiffer, *Introduction to General Relativity*, McGraw-Hill, New York, 1965.

[12] J. Foster and J. D. Nightingale, *A Short Course in General Relativity*, Springer-Verlag, New York, 1995.

[13] E. F. Taylor and J. A. Wheeler, *Exploring Black Holes*, Addison Wesley Longman, New York, 2000.

Index